T0139883

Advances in Intelligent Systems and Computing

Volume 974

The series "Advances in Intelligent Systems and Computing" contains publications on theory, applications, and design methods of Intelligent Systems and Intelligent Computing. Virtually all disciplines such as engineering, natural sciences, computer and information science, ICT, economics, business, e-commerce, environment, healthcare, life science are covered. The list of topics spans all the areas of modern intelligent systems and computing such as: computational intelligence, soft computing including neural networks, fuzzy systems, evolutionary computing and the fusion of these paradigms, social intelligence, ambient intelligence, computational neuroscience, artificial life, virtual worlds and society, cognitive science and systems, Perception and Vision, DNA and immune based systems, self-organizing and adaptive systems, e-Learning and teaching, human-centered and human-centric computing, recommender systems, intelligent control, robotics and mechatronics including human-machine teaming, knowledge-based paradigms, learning paradigms, machine ethics, intelligent data analysis, knowledge management, intelligent agents, intelligent decision making and support, intelligent network security, trust management, interactive entertainment, Web intelligence and multimedia.

The publications within "Advances in Intelligent Systems and Computing" are primarily proceedings of important conferences, symposia and congresses. They cover significant recent developments in the field, both of a foundational and applicable character. An important characteristic feature of the series is the short publication time and world-wide distribution. This permits a rapid and broad dissemination of research results.

**** Indexing: The books of this series are submitted to ISI Proceedings, EI-Compendex, DBLP, SCOPUS, Google Scholar and Springerlink ****

More information about this series at http://www.springer.com/series/11156

Amic G. Ho
Editor

Advances in Human Factors in Communication of Design

Proceedings of the AHFE 2019 International Conference on Human Factors in Communication of Design, July 24–28, 2019, Washington D.C., USA

Springer

Editor
Amic G. Ho
School of Arts and Social Sciences,
Creative Arts
Open University of Hong Kong,
Jubilee College
Ho Man Tin, Kowloon, Hong Kong

ISSN 2194-5357 ISSN 2194-5365 (electronic)
Advances in Intelligent Systems and Computing
ISBN 978-3-030-20499-0 ISBN 978-3-030-20500-3 (eBook)
https://doi.org/10.1007/978-3-030-20500-3

This Springer imprint is published by the registered company Springer Nature Switzerland AG
The registered company address is: Gewerbestrasse 11, 6330 Cham, Switzerland

Advances in Human Factors and Ergonomics 2019

AHFE 2019 Series Editors

Tareq Ahram, Florida, USA
Waldemar Karwowski, Florida, USA

10th International Conference on Applied Human Factors and Ergonomics and the Affiliated Conferences

Proceedings of the AHFE 2019 International Conference on Human Factors in Communication of Design, held on July 24–28, 2019, in Washington D.C., USA

Advances in Affective and Pleasurable Design	Shuichi Fukuda
Advances in Neuroergonomics and Cognitive Engineering	Hasan Ayaz
Advances in Design for Inclusion	Giuseppe Di Bucchianico
Advances in Ergonomics in Design	Francisco Rebelo and Marcelo M. Soares
Advances in Human Error, Reliability, Resilience, and Performance	Ronald L. Boring
Advances in Human Factors and Ergonomics in Healthcare and Medical Devices	Nancy J. Lightner and Jay Kalra
Advances in Human Factors and Simulation	Daniel N. Cassenti
Advances in Human Factors and Systems Interaction	Isabel L. Nunes
Advances in Human Factors in Cybersecurity	Tareq Ahram and Waldemar Karwowski
Advances in Human Factors, Business Management and Leadership	Jussi Ilari Kantola and Salman Nazir
Advances in Human Factors in Robots and Unmanned Systems	Jessie Chen
Advances in Human Factors in Training, Education, and Learning Sciences	Waldemar Karwowski, Tareq Ahram and Salman Nazir
Advances in Human Factors of Transportation	Neville Stanton

(continued)

(continued)

Advances in Artificial Intelligence, Software and Systems Engineering	Tareq Ahram
Advances in Human Factors in Architecture, Sustainable Urban Planning and Infrastructure	Jerzy Charytonowicz and Christianne Falcão
Advances in Physical Ergonomics and Human Factors	Ravindra S. Goonetilleke and Waldemar Karwowski
Advances in Interdisciplinary Practice in Industrial Design	Cliff Sungsoo Shin
Advances in Safety Management and Human Factors	Pedro M. Arezes
Advances in Social and Occupational Ergonomics	Richard H. M. Goossens and Atsuo Murata
Advances in Manufacturing, Production Management and Process Control	Waldemar Karwowski, Stefan Trzcielinski and Beata Mrugalska
Advances in Usability and User Experience	Tareq Ahram and Christianne Falcão
Advances in Human Factors in Wearable Technologies and Game Design	Tareq Ahram
Advances in Human Factors in Communication of Design	Amic G. Ho
Advances in Additive Manufacturing, Modeling Systems and 3D Prototyping	Massimo Di Nicolantonio, Emilio Rossi and Thomas Alexander

Preface

Communication of design is concerned with understanding users, creating communication, and engaging experiences. Users expect communication to be an exciting and interactive experience. Therefore, to provide such enjoyable and stimulating experiences, researchers in the discipline of communication design have identified that human factors try to enrich communication and optimize design experience. Topics covered in this book report on humanistic approaches, including the roles of human needs, emotions, thoughts, and actions, and have revealed more innovative approaches based on visuals. Furthermore, these types of connections influenced users' experiences in design consumption. Such experiences generally exerted a considerable effect on users' satisfaction toward the designs. Considering human factors in the communication of design enables designers to be actively connected with human needs.

The exploration of human factors and design in the past decades is an appropriate and valuable opportunity to enrich and strengthen the field of communication design. The book is organized into two sections that focus on the following subject matters: Communication of Design Methodologies and Communication of Design Applications.

This book will be of special value to a large variety of professionals, researchers, and students in the broad field of communication of design and human–computer interaction, who are interested in feedback of devices' interfaces, user-centered design, and design for special populations, particularly the elderly. The book is organized into four sections:

Section 1 Communication Design and Practices
Section 2 Design in Advertising and Media Communication
Section 3 Creative Arts
Section 4 Communication in Design

Each section contains research papers that have been reviewed by members of the International Editorial Board. Our sincere thanks and appreciation to the board members as listed below:

Seung Hyun Cha, Korea
E. Hung, Hong Kong
Kyle Kim, Korea
Sui Kwong (Sunny) Lam, Hong Kong
Kyu Ha Shim, USA
Jun Bum Shin, USA
Michael Siu, Hong Kong
Xiang Yang Xin, Macau
Jesvin Yeo, Singapore

We hope this book is informative, but even more—that it is thought provoking. We hope it inspires, leading the reader to contemplate other questions, applications, and potential solutions in creating good designs for all.

July 2019 Amic G. Ho

Contents

Communication Design and Practices

Nudge Users to Healthier Decisions: A Design Approach to Encounter Misinformation in Health Forums

Mahdi Ebnali[1(✉)] and Cyrus Kian[2]

[1] Applied Cognitive Engineering Lab, Industrial and System Engineering Department, University at Buffalo, Buffalo, NY, USA
mahdiebn@buffalo.edu
[2] Department of Information Sciences, Cornell University, Ithaca, NY, USA

Abstract. Health websites have increasingly become one of the main sources of health information for people around the world. Although health-seekers potentially benefit from these sources, misinformation in health forums significantly misleads people. We hypothesize that nudging users to think critically about the information might be an effective approach to encounter with effects of misinformation on their' health. To this end, we applied elaboration-likelihood-model (ELM) in web design to explore how central and peripheral cues influence on information evaluation and eye-movement data. The results show that providing central cues such as trusted links and discussion threads encourage the participants to look at the information with more critical thinking compared to the peripheral cues such as number of stars and likes. Similarly, fixation duration and time spent in each scenario confirmed that users engaged in higher level of cognitive processing and critical thinking when central cues are dominated ELM initiators.

Keywords: Health forums · Elaboration Likelihood Model (ELM) · Health misinformation · Eye tracking · Web design

1 Introduction

In the last decade, the internet has increasingly become one of the main sources of health matter information for people around the world. With the spreading of online communication, health-related websites have provided people with a massive amount of information. Easy access to the internet and rapid proliferation of health information in online Social Media encourage health seekers to consider the internet as a primary source of acquiring knowledge. Recent research shows that the majority of people who use the internet, search for information related to their health concerns [1]. These sources, especially health forums, offer support to health seekers by providing information and connecting with others in similar circumstances. Although health-seekers potentially benefit from these sources, misinformation in health forums significantly mislead people.

Misinformation is created and propagated in online networks because of the wide variability of members, sources credibility and gatekeepers' trustworthiness [2, 3].

A. G. Ho (Ed.): AHFE 2019, AISC 974, pp. 3–12, 2020.
https://doi.org/10.1007/978-3-030-20500-3_1

Validating the information on the online health communities is a challenging issue and pitfalls by users are far too often. Some data-driven approaches used real-time detection methods to encounter this challenge [4, 5]. However, because of the enormous amount of information and the semantic nature of the contents, the accuracy of detection drops in one side and computational costs hike in another side.

One other approach is persuading people to think critically about the messages in health forums and consequently, shift them toward healthier behaviors. Delving into users' decision-making in the online health world of mouth (WOM) and considering some design modalities might effectually guide users to make more rational decisions when faced with health-related contents posted by peers [6, 7]. However, directly asking users to think critically instead of blindly accepting or rejecting a message should be really annoying and frustrating while they are freely browsing in the net. In order to shift users' mind indirectly to critical thinking, we need to understand how decision-making mechanism is working in social situations. Health seekers make decisions to whether immediately accept/reject a message, even share it and act as a spreader, or critically think about it and explore for additional confirmatory information. This process plays a pivotal role in misinformation propagation and it's infecting impacts on individual and community [6].

Decision making and building attitudes are omnipresent in people daily activities and depend on the significance of consequence, it occupies different amount of mind processing capacities [8]. Health seekers are forced to consider important consequences of this judgment on their health, and it is expected that they critically investigate different aspects of their choices. However, social sciences studies have shown that sometimes people jump over this arduous step and simply build a positive or negative attitude toward a message, especially if this information is connected to their prior experience and belief [9–11]. Moreover, well-supported results in online social studies have indicated that beyond of the content of primary message, some secondary information such as the number of likes and stars substantially upsurge the validity and trustworthiness of messages [12–14]. [14], specifically, explored users' decision in health forums and revealed that users stated greater perceived credibility of the posts with high star rating in comparison to the posts with a low star grade.

Numerous line of studies has proposed theories and frameworks to expand our understanding of the persuasion process and how to estimate people reaction to different types of presentation. Elaboration Likelihood Model (ELM) is one of these theories that has been supported by notable experiential results [15]. This framework could assist us to comprehend the persuasive communication in health forums, how users form their attitude, and the extent they engage in elaboration or critical thinking.

ELM postulates that persuasion and building an attitude originate from how information passes through the central or peripheral route. Each of these routes differs in terms of the amount of thoughtful information processing or "elaboration" demanded of individual subjects. Elaboration is explained as effort people make to evaluate, remember, and change the attitude when they face with a piece of persuasive information. The level of elaboration then regulates which processing route a message might take: central route (high) processing or peripheral route (low) processing. Central route processing means users scrutinize the quality and strength of the information.

Contrarily, in peripheral route, people care more about superficial factors like source reputation, visual appeal, and presentation.

In online social research, numerous studies have evaluated how ELM model explain user behavior and attitude formation. [16], for example, investigated the persuasive effects of web personalizing features such as level of preference matching, recommendation set size, and sorting cue on customers' attitude. Similarly, [17] confirmed the explanatory power of ELM in influence and persuasion process in IT products. Other online experiment assessed how ELM elucidates attitude forming among online customers and reported that users with low skepticism were more intended to adopt peripheral cues [18]. Reversely, people with a high level of skepticism based their attitudes according to what they have already believed.

According to these findings, we hypothesized that if more central route initiators are provided in a health forum, people might be more persuaded to think critically about the post, and consequently build rational attitude toward the health message. In contrast, when peripheral cues are dominant initiators in a health website, people may think less critically about the information, and they are more likely to make heuristic decisions. Based on this hypothesis, two versions of a health forum with the same health contents (question and comment) were designed to evaluate whether users are persuaded to accept/reject information blindly or think critically about the information and look for more confirmatory cues.

As shown in Fig. 1, one design prototype (C-prototype) focuses on providing central cues and presents a thread of discussion and a trusted link with a small description related to the topic and comment. Another prototype (P-prototype) emphasis on peripheral cues and displays the number of likes and the star rating for the user who posted the comment. We theorized that the C-prototype increases the likelihood of higher information processing and increases the level of critical thinking and skepticism. Implanting seeds of skepticism in users' mind may prevent them from being influenced by misinformation. On the other hand, peripheral cues such as high star rating and the number of likes may escalate perceived credibility and quality of information because the information might pass through the peripheral processing route in ELM.

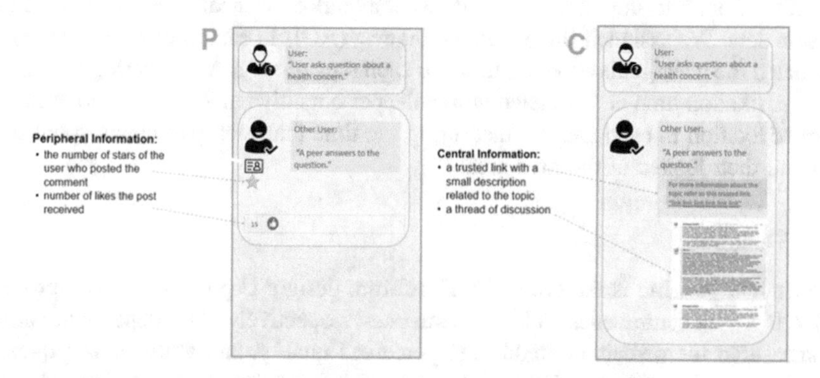

Fig. 1. Health forum prototypes with central (C-prototype) and peripheral cues (P-prototype)

2 Methods

2.1 Questionnaires

Studies in information literacy, marketing, and social sciences usually evaluate the information based on perceived credibility, perceived quality, and people' action and attitude [19–21]. Accordingly, a questionnaire was developed to measure the participants' attitude and how they evaluate the quality and credibility of the information provided in each prototype. Rating scale for these questions starts from −5 (indicates extremely disagree) continues to 0 (I am not sure), and ends to +5 (indicates extremely agree). We assume that participants who are persuaded to accept or reject a message are more likely to answer the questions close to two ends of rating spectrum (extremely agree or extremely disagree); and users who are nudged to think critically and look at the message skeptically, are more likely to select answers with more uncertainty (0 or close to 0).

Moreover, previous results indicated that people characteristics like prior attitude and level of critical thinking effect on their behavior in online contexts. Literature in social judgment and persuasion have shown that information is more likely to be accepted by people when it is consistent with their prior beliefs [9–11]. People assess the logical compatibility of the information with respects to their beliefs. Once a new piece of information has been accepted, it is highly resistant to change. To explore how these two factors effect on the results, we asked the participants to answer questions about "how they use health forums and online social media for their health concerns", "how they evaluate the quality and credibility of the information in online social environments", and "how they like to think about the new information critically".

2.2 Eye Tracking

In this study, using Tobii Pro Glasses 1, we recorded eye movement data such as fixation duration and saccade time to investigate participants' cognitive activity and visual attention. These parameters could screen how users looked at which part of the webpage, and how much time they spent in each section. Moreover, to visualize which part of the health forum attracted more visual attention, scanpath were generated using fixation data. We assigned three areas of interest (AOI) in each prototype (Fig. 3), and calculated the average fixation duration for each scenario and AOI, bearing in mind that a longer fixation time is associated with a deeper cognitive task. We also identified the order of fixation in each area by measuring the time that each participant spent in this area and then looked at the next area.

2.3 Variables

Independent variable is the type of health forum design: C-prototype and P-prototype with a focus on central cues and peripheral cues, respectively. As a dependent variable, we measured the perceived credibility, perceived quality, and attitude. The questionnaire contained two items about the quality of information, two questions about the credibility of information, and two last questions explored whether people act based on the provided content. Objective dependent variables include total time spent in each scenario, average fixation time, and order of fixation in AOIs. We used ANOVA to explore whether the type of web design and features impose significant changes in users' attitude and critical thinking.

2.4 Participants

Twenty-three participants (13 males, M age = 25, 10 females, M age = 26) participated in this experiment. The subjects ranged in age from 18 to 31 and reported that they had high levels of internet exploring experience. The recruiting flyers for the experiment were posted on the board of Shahrekord University and e-mail was sent to college students and the employees of the university. Forty-three people were interested in the experiment, but after the screening process, 23 participants were selected. The screening criteria included the level of their familiarity and activities in online contexts, and specifically health forums. People who did not have any engagement in health forums were excluded from the experiment.

2.5 Design of Experiment

This experiment was a within-subject study where the participants entered an experiment with eight scenarios, half in P-prototype and another half in C-prototype. As shown in Table 1, two of four scenarios in P-prototype showed a large number of like (32 likes) for the comment and highest star rating (5 stars) for the person who left the message. In other two scenarios, a few likes (3 likes) and lowest star rating (1 star) were shown. In C-prototype, the participants were presented long discussion threads (5 replies) in half of the scenarios and short discussion thread (1 reply) in another half. All scenarios in this group had a trusted link with a description of the health question. As a within-subject experiment, we had to avoid the effect of maturation; therefore, a counterbalanced design was considered in the experiment.

2.6 Procedure

Before starting the experiment, the participants were asked to read the informed consent form and sign it if they are agreed to contribute to the research. Then, they answered the demographic questions, items related to self-reported critical thinking, and familiarity with online health forums. Each subject needed to participate in the experiment with eight scenarios, equally from two common health problem scenarios: smoking cessation and weight losing. Each scenario consists a question- asking for a suggestion about smoking cessation or weight losing - and follows by other person's answer (comment). The participants required to read the question and the solution in the comment section. Finally, they should go to the next page and answer a questionnaire about the information evaluation.

In order to reduce the effect of comment content on users' attitude and persuasion, we provided an answer that is not realistic and common in each scenario. For example, the strange name of tablets and devices were suggested in the comment sections. Moreover, to control the effect words and sentences on the time spent in each scenario, we exactly used the same number of letters in question and comment sections. We conducted a pilot study with three users to assure they can understand the question and comment meanwhile they do not have any idea and prior attitude about the suggestion. In this case, instead of relying on the content, users have to explore for more peripheral and central cues provided in each prototype.

Table 1. Types of health forum prototype and scenarios - the number of like and stars: low = 3 likes, 1 star, high = 32 likes, 5 stars; length of discussion thread: short = 1 reply, long: 5 replies.

	Scenarios	Type of ELM cues	
P-prototype		Number of like and stars	
	Se1: weight losing	Low	High
	Se2: smoke cessation	Low	High
C-prototype		Length of the discussion thread	
	Se3: weight losing	Short	Long
	Se4: smoke cessation	Short	Long

3 Results and Discussion

Our data set consists of the qualitative evaluation of information and gaze data from 23 subjects, each participant in eight scenarios, using two prototypes of health forum. In the questionnaire data, we reverted all negative score to positive to avoid the effect of the sign of a number on the results. The Pair t-test and ANOVA test were run to evaluate how type of design effects on information evaluation and eye movement.

The questionnaire data disclosed that type of web design meaningfully effects on subjective evaluation of information and critical thinking. The participants assessed the information more skeptically in C-prototype compared to P-prototype ($p < 0.05$), even the same health information was presented. Figure 2 displays the difference of results separated by type of design, and three categories of information evaluation. Analysis of peripheral information shows that increasing the number of likes and stars, significantly amplified the perceived credibility and quality of information. In line with these results, earlier studies reported that the number of likes and shares considerably rise the validity and trustworthiness of messages [12–14].

Post hoc (Tukey) test reveals no statistically meaningful variance in information evaluation between P-prototype - with a low number of likes and stars - and all scenarios in C-prototype. Similarly, changing the length of discussion thread did not effect on people skepticism and attitude. It seems that a small number of likes and star rating instead of persuading people about the quality of information significantly reduces its credibility and quality. Likewise, [22] discussed that weak rating impairs users' persuasion and leads to a negative attitude toward the message. The topic of scenarios did not affect the information evaluation. Additionally, in contrast to [18], our results show that the participants' characteristic such as the level of critical thinking and their attitude toward online information did not influence the perceived credibility and quality of information. As a possible reason for this finding, the comments were not among common and expectable suggestions. Therefore, it seems that people prior attitude and experience did not manipulate the results.

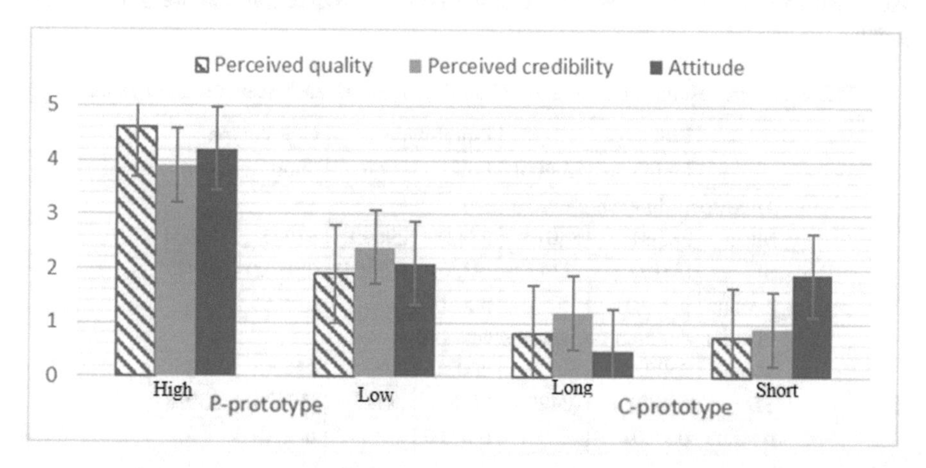

Fig. 2. Subjective evaluation of information and critical thinking categorized based on the type of health forum design

Analysis of the time spent in each scenario, shown in Table 2, separated based on the type of web design and features of each prototype. The participants spent longer time in C-prototype scenarios compared to P-prototype ($p < 0.01$, P-prototype: $M_{fixation\text{-}duration} = 28560$ ms, $SD_{fixation\text{-}duration} = 8301$, C-prototype: $M_{fixation\text{-}duration} = 84013$, $SD_{fixation\text{-}duration} = 21470$). In addition, the results suggest that the length of the discussion thread in a C-prototype scenario positively associated with the time spent this scenario. Similarly, the average fixation time in each scenario revealed that providing central cues such as a trusted link and discussion thread in a page meaningfully increases the fixation time compared to the scenarios with peripheral cues ($p < 0.01$, P-prototype: $M_{fixation\text{-}duration} = 25328$ ms, $SD_{fixation\text{-}duration} = 6423$, C-prototype: $M_{fixation\text{-}duration} = 69756$, $SD_{fixation\text{-}duration} = 24203$). Moreover, in C-prototype scenarios, increasing the length of the discussion thread leads to longer fixation time in each scenario ($p < 0.05$, short: $M_{fixation\text{-}duration} = 42500$ ms, $SD_{fixation\text{-}duration} = 19434$, long: $M_{fixation\text{-}duration} = 97012$, $SD_{fixation\text{-}duration} = 35528$). However, in P-prototype, changes in the number of likes and stars did not effect on fixation duration. In line with previous studies, this finding suggests that central cues triggered more visual and cognitive attention.

Three AOIs were defined in each prototype and are illustrated in Fig. 3 using yellow boxes. These AOIs in C-prototype include the areas for "comment", "trusted link", and "discussion thread". In P-prototype, AOIs are the areas for "comment", "likes", and "stars". Scanpath analysis (Fig. 3-b) shows an obvious difference between visual scanning pattern of the users in C-prototype and P-prototype scenarios. This analysis demonstrates that "discussion threads" triggered the participants' visual attention more than the "trusted link". Moreover, as shown in Fig. 3-b, similar results are extracted from the order of AOIs scanning. First, the users checked the comments, then discussion thread, and finally the trusted link. In P-prototype scenarios, AIOs of "like" and "stars" received similar visual attention. The general order of AIOs, show that people first looked at the number of likes then explored for more peripheral cues such as user's stars. No significant difference was observed in scanpath and order of AOIs with respect to the type of scenarios (smoking cessation and weight loss) and gender.

Table 2. The results for data analysis on time spent in each scenario and eye data.

	Prototype		P-prototype		C-prototype	
	Peripheral	Central	Low	High	Short	Long
Time spent in each scenario (ms)						
Mean	28560	84013	27421	29699	61075	106951
SD	(8301)	(21470)	(9326)	(12404)	(29842)	(41940)
p value	p < .01		p > .05		p < .05	
Fixation duration on whole scenario (ms)						
Mean	25328	69756	26021	24635	42500	97012
SD	(6423)	(24203)	(7366)	(8523)	(19434)	(35528)
p value	p < .01		p > .05		p < .01	

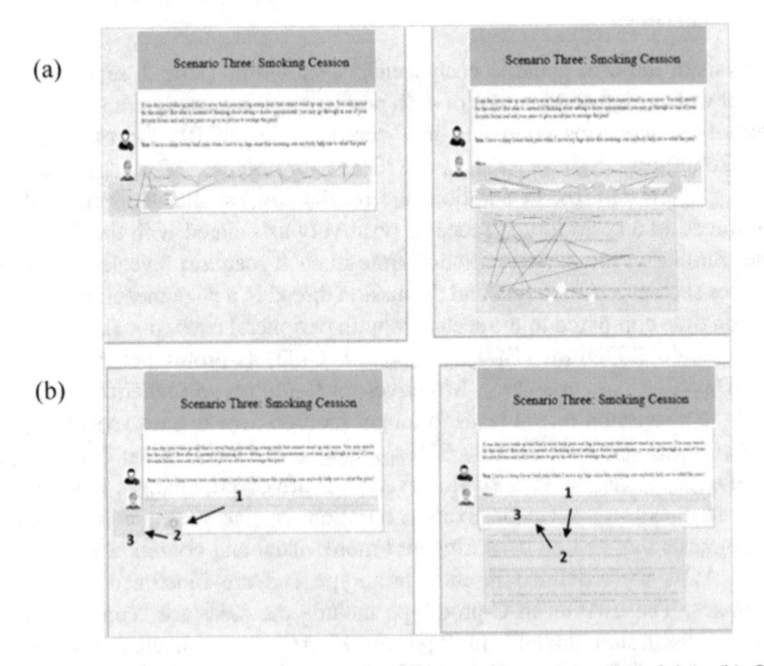

Fig. 3. (a) Users' scanpath in P-prototype (top-left) and C-prototype (top-right), (b) Order of scan in three AIOs in P-prototype (down-left) and C-prototype (down-right)

In general, eye movement data suggest that central cues in the health forum triggered more cognitive processing. These results can be explained by prior research findings. According to ELM, central cues causes that information passes through the central route and consume more cognitive resources. In line with our finding, [23] proposed that longer fixation time of users in online environments is linked with high level of cognitive processing. Furthermore, Huitt is his review [24], discussed that higher cognitive processing positively is associated with more critical thinking. Consistent with these findings, central cues like a trusted link and a tread of discussion successfully nudge users to more level of critically thinking, and ideally avoid health seekers to be affected by health misinformation.

4 Conclusion

The findings of this research indicate that application of ELM in designing health forum significantly effects on the participants' subjective evaluation of information and their critically thinking. The result of questionnaire revealed that providing central cues such a trusted link and discussion thread encourage the participants to look at the information with more skepticism and critical thinking compared to the peripheral cues such as the number of stars and likes. Similarly, fixation duration and time spent in each scenario confirmed that users engage in a higher level of cognitive processing and critical thinking when central cues are dominated ELM initiators. These findings suggest that designer and vendors provide more central cues in health websites and forums to encourage users to think critically about new information.

References

1. Freeman, J.L., et al.: How adolescents search for and appraise online health information: a systematic review. J. Pediatr. **195**, 244–255 (2018)
2. McNab, C.: What social media offers to health professionals and citizens. SciELO Public Health (2009)
3. Moorhead, S.A., et al.: A new dimension of health care: systematic review of the uses, benefits, and limitations of social media for health communication. J. Med. Internet Res. **15** (4), e85 (2013)
4. Baeth, M.J., Aktas, M.S.: Detecting misinformation in social networks using provenance data. Concurrency Comput.: Pract. Exp. **31**(3), e4793 (2019)
5. Kinsora, A., et al.: Creating a labeled dataset for medical misinformation in health forums. In: 2017 IEEE International Conference on Healthcare Informatics (ICHI). IEEE (2017)
6. Morahan-Martin, J., Anderson, C.D.: Information and misinformation online: recommendations for facilitating accurate mental health information retrieval and evaluation. CyberPsychol. Behav. **3**(5), 731–746 (2000)
7. Wilson, B.J.: Designing media messages about health and nutrition: what strategies are most effective? J. Nutr. Educ. Behav. **39**(2), S13–S19 (2007)
8. Darley, W.K., Blankson, C., Luethge, D.J.: Toward an integrated framework for online consumer behavior and decision making process: a review. Psychol. Mark. **27**(2), 94–116 (2010)

9. Del Vicario, M., et al.: The spreading of misinformation online. Proc. Natl. Acad. Sci. **113** (3), 554–559 (2016)
10. Sung, K.H., Lee, M.J.: Do online comments influence the public's attitudes toward an organization? Effects of online comments based on individuals' prior attitudes. J. Psychol. **149**(4), 325–338 (2015)
11. DiFonzo, N., Bordia, P.: Rumor Psychology: Social and Organizational Approaches, vol. 1. American Psychological Association, Washington, DC (2007)
12. Metzger, M.J.: Making sense of credibility on the web: models for evaluating online information and recommendations for future research. J. Am. Soc. Inform. Sci. Technol. **58** (13), 2078–2091 (2007)
13. Morahan-Martin, J.M.: How internet users find, evaluate, and use online health information: a cross-cultural review. CyberPsychol. Behav. **7**(5), 497–510 (2004)
14. Sauls, M.E.: Perceived Credibility of Information on Internet Health Forums. Clemson University (2018)
15. Petty, R.E., Cacioppo, J.T.: The elaboration likelihood model of persuasion. In: Communication and Persuasion, pp. 1–24. Springer (1986)
16. Tam, K.Y., Ho, S.Y.: Web personalization as a persuasion strategy: an elaboration likelihood model perspective. Inf. Syst. Res. **16**(3), 271–291 (2005)
17. Bhattacherjee, A., Sanford, C.: Influence processes for information technology acceptance: an elaboration likelihood model. MIS Q. **30**, 805–825 (2006)
18. Sher, P.J., Lee, S.-H.: Consumer skepticism and online reviews: an elaboration likelihood model perspective. Soc. Behav. Pers.: Int. J. **37**(1), 137–143 (2009)
19. Eastin, M.S.: Credibility assessments of online health information: the effects of source expertise and knowledge of content. J. Comput. Mediated Commun. **6**(4), JCMC643 (2001)
20. Ayeh, J.K., Au, N., Law, R.: "Do we believe in TripAdvisor?" Examining credibility perceptions and online travelers' attitude toward using user-generated content. J. Travel Res. **52**(4), 437–452 (2013)
21. Hu, Y., Shyam Sundar, S.: Effects of online health sources on credibility and behavioral intentions. Commun. Res. **37**(1), 105–132 (2010)
22. Cheng, V.T., Loi, M.K.: Handling negative online customer reviews: the effects of elaboration likelihood model and distributive justice. J. Travel Tourism Mark. **31**(1), 1–15 (2014)
23. Pan, B., et al: The determinants of web page viewing behavior: an eye-tracking study. In: Proceedings of the 2004 Symposium on Eye Tracking Research & Applications. ACM (2004)
24. Huitt, W.: Critical thinking: an overview. Educ. Psychol. Interact. **3** (1998)

How Creative Mindset Operates with Respect to Creative Performance: Pedagogical Factors that Ignite Creative Mindset in Design Education

Joungyun Choi(✉)

College of Environmental Science and Forestry, State University of New York, 1 Forestry Dr., Syracuse, NY 13210, USA
jchoi@esf.edu

Abstract. This research investigated college design students' mindset for the "decision to be creative" when they have a design project in front of them; moreover, this research investigated factors that influence students' creative mindset and how educators enhance creative mindset to students. In order to address these purposes, a qualitative triangulation was used. The major findings of this study provide a number of important implications for improving pedagogical strategy, utilizing the concept of the creative mindset within a design education context. Moreover, this study provides directions for future research regarding the creative mindset in design education and suggestions for expanding this study to contribute to other disciplines that is needed creative thinking.

Keywords: Creative mindset · Design education · Higher education · Pedagogic strategies · Decision to be creative · Design process

1 Introduction

Researching people's mindset in terms of creativity is important because their mindset play a critical role in the affection of their positive attitude. Creative performance usually starts with a positive attitude toward the creative process. A mindset is a particular way of thinking, tendency, or habit. People tend to start something from their own perspective and come up with ideas that are novel and useful in some way. This tendency to develop creative work originates mostly from the decisions that people make. People have the willingness to discover ambiguity, overcome obstacles, and find ways to do the necessary work before executing creative performance.

According to the investment theory of creativity, creativity is a decision similar to that of investment:

…to be creative is to invest one's abilities and efforts in ideas that are novel and of high quality, and one must, like any good investor, *first decide to* generate new ideas, analyze these ideas and sell the ideas to others… People are not born creative or uncreative. Rather, they learn and develop a set of attitudes toward life that characterize those who are willing to go their own way [1].

A. G. Ho (Ed.): AHFE 2019, AISC 974, pp. 13–22, 2020.
https://doi.org/10.1007/978-3-030-20500-3_2

The main purpose of this research is to gain knowledge regarding the relationship between design students' creative mindset (decision be creative) and their attitudes that motivate them to engage in design work creatively.

I believe that students' positive attitudes, such as willingness and curiosity, can be explained by intrinsic motivation. For example, curiosity is defined as "a desire to know, to see, or to experience something that motivates exploratory behavior directed towards the acquisition of new information" [2]. The drive theory of curiosity refers to a person's cognitive and perceptual coherence, which can become disrupted by novel, complex, or ambiguous stimuli. By learning and gathering new information about related stimuli, it is possible to restore cognitive and perceptual coherence [2]. Put differently, curiosity toward stimuli facilitates one's willingness to *decide to act.*

This conception of intrinsic motivation can be applied to the creativity domain in design education. Indeed, intrinsic motivation can explain what a student's driving force is in being creative, and how he or she uses it for his or her design in order to be more creative. According to Amabile [3], intrinsic motivation arises from the individual's perceived value of engaging in the work itself. People seldom work creatively without really enjoying it versus obtaining possible rewards. Intrinsic motivation makes people become interested in something that they want to know more about; consequently, it makes them more likely to act upon their thoughts. A creative process starts with finding and formulating a problem [4]. In other words, people should be willing to solve problems and should be curious about such problem, as well as the solutions. By using these overarching ideas of investment theory and intrinsic motivation, I define the creative mindset as *the decision to be creative*.

Research involving students' mindsets in regard to the decision to be creative in an educational context is very important because one's attitude toward life reflects his or her willingness to follow his or her own way, which can be learned and developed. Generally, spontaneous attitudes appear in the process of creative work; furthermore, it has been found that people do their most creative work when their motivation is intrinsic rather than extrinsic. In this case, people care about what they are doing, and not just about what they will eventually end up with. In other words, they have a learning goal orientation: involves mastering new things, not a performance goal orientation: based on measuring ability [3, 5]. The learning goal orientation concept is subset of a growth mindset theory. This theory explains people's beliefs regarding their psychological traits and abilities, which plays an important role in influencing their own motivation and behavior [5]. People who believe that their abilities are able to develop and change. The current study aims to establish the fact that the growth mindset concept can be applied to the creative mindset in design education. As a result, design students can develop their creativity.

2 Method

Research focused on *design students*' decisions and implicit beliefs about creativity is very rare. Although this study is based on existing theories of creativity, mindsets, and student behavior, the main purpose of this study is to gain knowledge in regard to examine creative design students' states of mind (thinking styles), the creative mindset, and design students' acts of creative performance. An additional aim of this study is to establish factors that motivate students' creative mindset in design educational context. Thus, grounded theory is a part of the research tools in this study, given that this research seeks and conceptualizes latent design students' thinking patterns, relating them to growth mindset concept. The main challenge of this study is related to the nature of the educational environment, as well as to students' individual thinking styles and mindsets on creativity and their performance. These factors are complex: they represent situational change processes, and they are multifaceted and multi-actor in nature. Thus, qualitative triangulation was employed.

2.1 Participants and Setting

Developing the creative mindset is an important value of an educational goal, particularly in design education. For this reason, design students were selected as the participants of this study. Because a training period was involved in using the design tools, and because experience in doing design projects led to a greater likelihood of affecting students' performance and design outcomes, undergraduate design students were the unit of analysis for this study. There were three phases in the process of the study, which were as follows: a growth mindset (toward creativity) test, direct in-depth interviews, and observations.

2.2 Growth Mindset Test

Using existing measurement of creative growth mindset [6–8], the test was conducted to identify students with a high growth mindset toward creativity. The sample for the test consisted 179 students who agreed to participate in this research. The students were enrolled in a variety of design subject matter tracks, such as graphic design, architecture, interior design, product design, and apparel design in a couple of different major universities located in the US Midwest. These students were selected from two introductory-level design classes and two advanced-level design classes.

2.3 In-depth Interview

When the test was completed, 12 participants with a high growth mindset were selected to continue to the interview phase. A qualitative data collection method was used with in-depth interviews consisting of open-ended questions to identify participants' implicit beliefs about creative mindsets (decisions to be (more) creative) and factors to enhance their creative mindset. The interviews were designed to collect descriptive data in the participants' own words and to develop an understanding with regard to participants' opinions and experiences. The interviews took approximately one hour each. All interviews were recorded and transcribed; moreover, the interview situations were documented in field notes. The data were analyzed through an inductive content analysis, in which "themes and categories emerge from the data through the researcher's careful examination and constant comparison" [9].

2.4 Observations

Throughout the entire research process, participant observations were conducted to examine their behavior/performance and outcomes of the creative design process. To enhance the validity of the observation data, the observation took four months (one semester) in two different design studio classes that were selected for phase 1 (the tests). Two classes were visited within two time periods (4 h) a week during the entire semester. The observations were made when the participants were given a new project and time to work on their own. In this way, the relationship between participants' creative mindsets (decision to be creative) and creativity (developing their project) in the design process could have been observed through their behavior. To analyze the field notes, (1) the collected data were organized into a narrative format of a day; (2) the narrative format of the information was organized, according to the outline of the research questions; and (3) a deductive content analysis was used, which "starts with the counting of words or manifest content, then extends the analysis to include latent meanings and themes" [9]. The outlined text information was analyzed to determine the frequency of the contents (students' habits of action). This information was later compared to the research objectives and interview findings to match with students' performance and the creative mindset.

3 Results

1. Making an effort to have new experiences and to get external stimuli helps one's motivation to be (more) creative. 100% of the students reported that creativity more or less depends on effort. Ten out of the 12 participants actually used the word "effort" in their interviews, and two participants gave examples to show how they make an effort to be more creative. Participants make an effort to have new experiences, since they believe that a creative outcome is based on an individual's experience. Participants' motivation to develop more creative ideas comes from external stimuli (experience). During the interviews, all of the participants shared their experiences. They said that these experiences triggered their motivation to engage in creative activities. The participants below said that they make an effort to expose themselves to new environments, such as going out to see nature, museums, traveling, meeting friends, and getting a new hobby. They also try to observe their surroundings from different perspectives and from breaking things down and reconnecting them in order to redefine something new. Participant 11 said that she believes an individual's experiences shapes one's creativity, especially experiences involving failure stimulate her to become more creative.

> Participant 1: I think I try to make quite a bit of *effort in being more creative*. When I feel that I need new ideas, I go out to a new environment. I tend to just take inspiration from anything I see and look at it from a design perspective. Nature, design websites, furniture, architecture, etc. and just breaking down its form, color, balance, shape, etc. I think noticing these details helps me be more creative in my designs.

Participant 2: … The *effort helps me be more creative* because I am actively observing other objects and trying to find ways in which I can apply that to my designs. I definitely think creativity levels can differ, based on effort in trying to be more creative. I think lesser creative people just take art and design as it is and don't try to find creative solutions or innovations.
Participant 4: I have noticed that going to museums and places where art and design live is something that always inspires new ideas and projects for myself. *Making the effort to just surround yourself with others works and sometimes gives me the push to start something new.*
Participant 6: I think *there is always a bit of effort needed when starting a new hobby*, but nothing over the top. I usually do low-level research to make sure I have a comprehensive idea of what something will take, and then I do it on my own. The process of figuring it out is where the creativity is reignited…making the hobby or new activity my own.
Participant 10: I naturally take *my effort to find good ideas in my everyday lives.* I observe and find things to connect to each other and mix them together to create something new. It happens when I walk, meet my friends, travel, and even lie down to sleep. I think that my creativity is based on my research. All of my experience can be research because I have a habit of journaling what I feel, see, hear, and read.
Participant 11: The ability to explore creative ideas depends on what kind of experience you have had. *Your experience shapes your creativity*…In my case, creative ideas come from my experts and things that I have already experienced, especially a failed experience. Challenges stimulate my creativity… It should be fueled by curiosity and a drive to be experimental and to try to do new things—either new to yourself or to the world. Even though it's not easy, trying to embrace failure is another part. Those who are innovative often fail the first, second, third (and so on) time.

Participant 5 makes an effort with a little different perspective. This participant said that he has high creativity, but sometimes external factors, such as negative habits, hinder his creativity. Thus, he makes an effort to overcome these negative habits and temptations to help himself become more creative.

Participant 5: I seek out other people's perspectives on what I'm working on and remove myself from digital temptations. If you have negative habits that disrupt your flow, making the effort to overcome them can help you be "more" creative… Overall, *I think that creativity can be less or more, depending on your effort.*

Participant 7 did not mention the word "effort" in the interview. Rather, she said that the activities in which she engages to maintain her own creative process mainly involve looking for external stimuli, such as critiques, attending collaborative events, and image-based research from social media sites.

Participant 7: I consciously (and sometimes unconsciously) search for inspiration in my environment. I also love all of those image-based social media sites like Pinterest, Behance and Dribble. I can't say no to looking at pretty things, but those sites also come in handy for inspiration searching or finding new styles to try. Going through all of the studio courses and the critiques with that have helped my creativity immensely. It helps to hear other peoples' opinions, plus learning from instructors. You also learn how to expand and hone in on your own creative process. I also enjoy attending collaborative events like workshops and conferences. I've have learned so much from those, and it has maintained my curiosity.

2. *Education plays an important role for students in having a creative mindset (real-world problem-based learning, opportunities to support peers' creativity).* The participants said that studio classes motivate their creative mindsets. Because a studio class mainly focuses on experience-based learning (creating things with their performance), it usually leads to their spontaneous project process. They plan, research, ideate, sketch, and make their projects. This process makes them think creatively. Ninety-two percent of the participants shared their experiences from design studio classes, and they are mostly motivated by the types of design projects.

The first type of project is one that has clear and specific goals to apply to a real situation. This type of project motivates students to actively engage in the project. This positive attitude enhances their creativity. Participant 5 discussed a project about building environmental graphics and a wayfinding system. This project was intended to create cost-effective and visually effective solutions for an inside space in a design department building that the participants use every day.

Participant 5: We redesigned a wayfinding system for a space in the building where we work every day. The building is an outdated old facility. My research and design could be a proposal for redesigning the building later. It makes me excited to do the project. I did a survey with the students and professors who use this building to know their needs. I felt that this process motivated me to be more creative because my design could reflect people's needs, as well, and has the potential to be used to improve our environment.

The second type of project is an assignment with an interaction between teaching and learning. This was a great opportunity to hear unexpected stories about interactions between teaching and learning. The participants said that interactive support is very helpful in being creative. Motivation to be creative comes not only from an instructor's support (lectures, feedback), but also the participants' own support of others.

Participant 6: …One of the team projects was to conduct research on an assigned historic era from Design History. Our team had to create presentation material containing many visual references and historic information. And then we had to give a lecture to the class using the material. After we finished the lecture, we did a workshop for other students to create artwork using an art style from the history lecture. Our charge for the workshop was to encourage students to make artwork "creatively." It was interesting to think about strategies that we can employ to help someone be creative. And I was fascinated to see the growth of the students… I was able to expand my thinking and figure out how to make their artwork better, which I think was really hard. But I think that really helped. I think that it improved my creativity.

The third type of project is a project requiring constructive critique, which involves specified instructions on how to give feedback. The participants reported that the constructive critique process makes them have an open mind in terms of accepting others' critiques. Also, it gives the participants confidence to give their feedback to others, since there are guidelines. At the same time, there is a chance for the participants to think about their own design. In this way, they are positively challenged by the critique process. Students are encouraged to be more creative when they have chances and options.

> Participant 11: Before I had the critique, I didn't even know how to say, or what to say to others' work. I sometimes said, "I like the color." "I like the way it is used in your design"... But I know that this can't help others develop their designs. We had a constructive critique for our 4th project. We were in groups of four. Our group had to give feedback to the whole class. The professor gave us a sheet with a list containing a bunch of items. The items tell where and how to look at others' designs and ways to give a feedback (at least one suggestion to improve the design, and at least one comment of what and why it is good). At first, our team discussed the design based on the list on the feedback sheet, wrote our comments on the sheet, and moved on to the next design. Once we finished the critique and came back to my design, my feedback sheet was full of helpful comments. This whole process made me think about how to make others' work better, and this thinking process gave me a chance to think about how my design can be better. I actually came up with a better and pretty creative idea for my design while I was doing this critique.

4 Discussion

During the interviews with the high creative growth mindset students, they shared their experiences of when they were motivated and felt encouragement to engage in creative performance. The most common aspect of motivating their creative mindset was an instructional activity and/or experience that encouraged them to have an open mind and to be intrinsically motivated. learning is a process by which students engage in activities (actively doing something), such as discussion, participation, hands-on workshops, or problem-solving that promote students' interest, enjoyment, and curiosity in their learning.

1. Active learning focused on what they "actively do"

In the interviews, common opinions consistent with active learning were found from the study participants. The participants reported that creativity is an activity intended to find new, innovative challenges that change a situation for the better. They also reported that creativity is not merely about a new product (outcome); rather, creativity is a process by which a person comes up with something new. They said that a design studio class, as opposed to a large lecture class, motivates their creative mindset. Because a studio class is mainly focused on creating things using the design process, it usually leads to their spontaneous activity to create outcomes. They plan, research, ideate, sketch, and make their projects. This process makes them think creatively. Basically, all of the participants have had positive experiences in design that required hands-on skills. This made them more likely to consider choosing design as their major. Due to this characteristic of design students, a course with opportunities to *do* practical projects and *participate* in real-life experiences will be more likely to motivate their creativity. Participating in real-life experience through design works, such as real-world problem-based projects and community-based work, would allow students to generate ideas and would encourage them to move forward in their creative projects. This kind of course, such as a design studio course that adopts a real-world problem-based project, requires a step-by-step process to solve problems. A creative mindset is involved during this process. The students said that every single time a new project was in front of them, they asked themselves how they could execute the project more creatively. Also, the active learning process makes students more likely to spontaneously plan and lead their projects so that their choices are applied at every step.

2. Opportunity to support a peer's creativity

An interaction between teaching and learning fosters a creative mindset. The meaning of "teaching" here indicates helping and supporting someone to be creative, while "learning" indicates receiving someone's support and encouragement in order to be creative. The participants said that a peer's engagement and encouragement, such as feedback, less negative criticism, more acceptance of diversity, and allowance for failure all motivate them to be more creative. Moreover, while they attempt to figure out how to make others' projects better, it enables them to expand their own thinking. Thus, the interaction between teaching and learning positively impacts an individual's creative mindset and provides them with an opportunity to support their peers' creativity while enhancing their own.

3. Relationship between external stimulation and intrinsic motivation

Clear guidelines (on how to give feedback) provide students with specific goals to develop their work and open their minds to accept others' critiques. Based on the participants' experiences, receiving only praise for their work and having the ability to engage in the design process do not foster motivation or lead to creative accomplishments. They were able to give the most effective and helpful feedback when they have guidelines of where to focus on others' work, what to comment on, and how to make others' work better. This way of giving comments encourages students who get feedback to be more creative (willing to develop).

Intrinsic motivation, which is strongly related with a creative mindset, arises from these external stimuli, such as teaching and learning relations, unrelated events, loose but constructive ways of learning, and unconventional ways of approaching problems. Students said that a loose, but constructive learning environment gives them the freedom to think about and do things; in this way, they are positively challenged. They are encouraged to be more creative when they have chances and options. Using unconventional ways of approaching problems is fun and enjoyable for them. These students feel that their creativity is tapped when they are involved in unstructured situations. They believe that the most creative ideas come from unrelated events. From students' reports of unrelated events, it is evident that they develop creative ideas when they are engaged in something different from the problem that they need to solve (e.g., talking with friends about a TV show, driving a car while traveling, daydreaming, eating food, etc.).

5 Conclusion

The major finding of the study could provide a number of important ways to improve pedagogical strategy, utilizing the concept of the growth mindset within a design education context. First, the factors that contribute to a student's creative mindset will contribute to developing a better design curriculum by integrating real-world problem-based projects and community-based works. Second, instructional factors can help educators set goals for class activities, such as giving students an opportunity to support their peers' creativity and setting constructive guidelines for critiques in the design process. Third, knowledge about the relationship between the creative mindset and the concept of the growth mindset will help educators create an instructional environment and situation that can ignite students' creative mindsets, such as unconventional ways of approaching design problems and introducing unrelated events. Particularly, creating a good atmosphere for the first impression of a project is important in leading students' creative mindsets, since the creative mindset involves students' attitudes toward the design project at the beginning stage. Also, the growth mindset is always involved in the process of creative thinking. As a result, educators can take advantage of situations in which students might express unexpected and surprising ideas. From this situation, educators praise the process and effort of creative thinking more than the outcome/creative ability, which is very important in encouraging students' intrinsic motivation.

Recommendation for future research is that interdisciplinary collaborations could be made, since the result could be implemented in other disciplines that require high creativity or creative design thinking, such as the sciences, applied economics, and business in higher education. Furthermore, the concept from this study could be undertaken among professional designers, who are experiencing different states of mind in different types of projects to see how these professional designers maintain their creative mindsets. Also, it would be interesting if this study could be broadened to the eminent creativity area. For example, a similar study using the same criteria may be undertaken among other professionals (scientists, authors, entrepreneurs, etc.) with significant achievements, such as Nobel Prize recipients or those who have contributed to major transformations within their fields.

References

1. Sternberg, R.J.: The Nature of Creativity. Contemporary Psychological Perspectives. Cambridge University Press, Cambridge (1988)
2. Litman, J.: Curiosity and the pleasures of learning: wanting and liking new information. Cogn. Emot. **19**(6), 793–814 (2005)
3. Amabile, T.: The Social Psychology of Creativity. Springer Series in Social Psychology. Springer, New York (1983)
4. Sawyer, R.: Exploring Creativity. The Science of Human Innovation, 2nd edn. Oxford University Press, New York (2012)
5. Dweck, C.S.: Self-theories: Their Role in Motivation, Personality, and Development. Psychology Press, Philadelphia (1999)

6. Hong, Y.Y., Chiu, C.Y., Dweck, C.S., Lin, D.M.S., Wan, W.: Implicit theories, attributions, and coping: a meaning system approach. J. Pers. Soc. Psychol. **77**(3), 588 (1999)
7. Kawowski, M.: Creative mindsets: measurement, correlates, consequences. Psychol. Aesthetics Creativity Arts **8**(1), 62 (2014)
8. Hass, R., Katz-Buonincontro, J., Reiter-Palmon, R., Tinio, P.: Disentangling creative mindsets from creative self-efficacy and creative identity: do people hold fixed and growth theories of creativity? Psychol. Aesthetics Creativity Arts **10**(4), 436–446 (2016)
9. Zhang, Y., Wildemuth, B.M.: Unstructured interviews. In: Wildenuth, B. (ed.) Applications of Social Research Methods to Questions in Information and Library Science, pp. 222–231. Libraries Unlimited, Westport (2009)

Exploration and Thinking on the Cultural Communication of Guangdong Museum

Xiaobao Yu(✉) and Ting Wang

Art & Design School, University of Shenzhen, Shenzhen, Guangdong, China
yxb99-1@126.com, 375352215@qq.com

Abstract. The cultural communication of traditional museums is centered on the exhibition of real objects. With the upgrading of cultural consumption, the public is no longer satisfied with the museum cultural consumption mode of reading-style. In the context of consumption of new culture, Guangdong Museum tries to explore a new form of museum culture communication by taking the cultural and creative product as a breakthrough. The way of cultural communication with the core of cultural and creative products is a multi-level mode of interactive communication, which is based on the cultural consumption demand as its starting point, the modern design as its means and the new media as its channel. This mode of communication breaks through the time and space boundaries of museum culture communication and makes the museum culture really enter the public life.

Keywords: Museums · Cultural communication · Creative products · Modern design

The Guangdong museum opened to the public in 1959. It is a first-class museum in China and one of the landmark cultural buildings in Guangzhou. Focusing on the history, culture, art and nature of Guangdong province, the museum shoulders the important functions of cultural communication and public service. The museum has rich collection resources. As of December 2016, the total collection of the museum has reached more than 172,700 pieces (sets), including 130,785 pieces (sets) of cultural relics and ancient books, and 41,960 pieces of natural specimens. As a place to collect natural and historical cultural heritage of Guangdong province, Guangdong museum has made some new explorations and reflections on how to meet the cultural needs of the public and realize the functions of culture education and communication.

1 The Traditional Mode of Cultural Communication in Museums

The cultural communication of traditional museums is centered on the collection of cultural relics, with little regard for the needs and feelings of the public. Therefore, the mode of cultural communication is mainly physical exhibition, which can be divided into basic exhibition and temporary exhibition. Basic exhibition mainly refers to permanent exhibition, while temporary exhibition can be divided into thematic activities

A. G. Ho (Ed.): AHFE 2019, AISC 974, pp. 23–31, 2020.
https://doi.org/10.1007/978-3-030-20500-3_3

and special exhibitions. The Guangdong museum has five permanent exhibitions. Taking Guangdong history and culture exhibition hall as an example, the exhibition contents are divided into four chapters with the local development history as the time clue. The collections on display are dominated by words and photographs, and tell the public about Lingnan's culture and history by means of sculpture, multimedia and restoration scenes. This kind of presentation is intuitive, but cultural communication focuses on one-way output and lacks cultural interaction.

Thematic activities and special exhibitions are also traditional ways of cultural communication in Guangdong museums. Guangdong museum held in November 2017 the "Mucha - European art nouveau treasure" for a period of more than three months temporary special exhibition, by showing the Czech famous painter Alphonse Mucha, general paintings and other new art movement painter's works, make domestic audience close to appreciate the Czech artist's paintings. At the same time, the museum also held a series of activities for visitors of different ages. Through these activities audiences can experience the charm of art.

The model of cultural communication with the exhibition of real objects as the core is a passive model of cultural appreciation. This mode of cultural communication is limited by the audience's own cultural literacy, and cultural communication only stays at the level of cultural popularization, lacking the depth and breadth of communication. With the development of society and the improvement of people's knowledge level, visitors are no longer satisfied with just appreciating the cultural relics and exhibits, but more want to explore the cultural connotation behind the cultural relics and exhibits, and even integrate their favorite culture into their own life.

2 Museum Culture Communication Mode with Cultural and Creative Products as the Core

Different from the traditional static display and graphic presentation of physical objects in museums, cultural and creative products are commodities bearing special cultural symbols, which realize socialized circulation through purchasing behavior. Their communication mode breaks through the limitation of space and time of traditional museums, and makes museum culture truly enter the public life. This mode is a multi-level interactive communication mode with museum cultural needs as the starting point, modern design as the means and new media as the channel.

2.1 Cultural Consumption Demand Is the Starting Point

(1) Use Value Requirements. The use value of cultural and creative products includes basic functions, quality, appearance, specification, safety and convenience, etc. The use value of products is an important measurement standard for consumers to purchase. Whether it is the spiritual level or the material level, only to meet certain needs of consumers can arouse people's interest in buying. After the author's investigation, according to the use value of the products, the cultural and creative products on the market mainly include the following categories (Table 1):

Table 1. The main categories or cultural and creative products on the market

Category	Variety
Stationery	Pen holder, bookmark, postcard, notebook, pen, ink
Household products	Sculpture decoration, refrigerator sticker, cup, pillow, mobile phone case
Clothing accessories	Scarves, clothes, shoes, bags, key chains, umbrellas
Jewelry watches	Necklaces, earrings, bracelets, brooches, watches, pocket watches
Holiday supplies	Greeting cards, gift wraps, bonus seals, couplets, moon cakes, etc.
Printed matter	Posters, books, brochures, guides, maps

(2) Memorial Needs. In addition to the functions of the products, many tourists choose to buy peripheral products of scenic spots to commemorate their experiences. In the survey on "the purpose of buying cultural and creative products", 60% of people buy products to collect and commemorate this journey, so commemorative will also become one of the reasons for viewers to buy cultural derivatives. Memorial is also reflected in the emotional design of products. In the process of people buying and using cultural and creative products in museums, they can inspire the related associations about the journey or culture, and generate the double experience of emotion and culture, so as to achieve the purpose of memorial (Table 2).

Table 2. Survey of "purchasing cultural and creative product purposes".

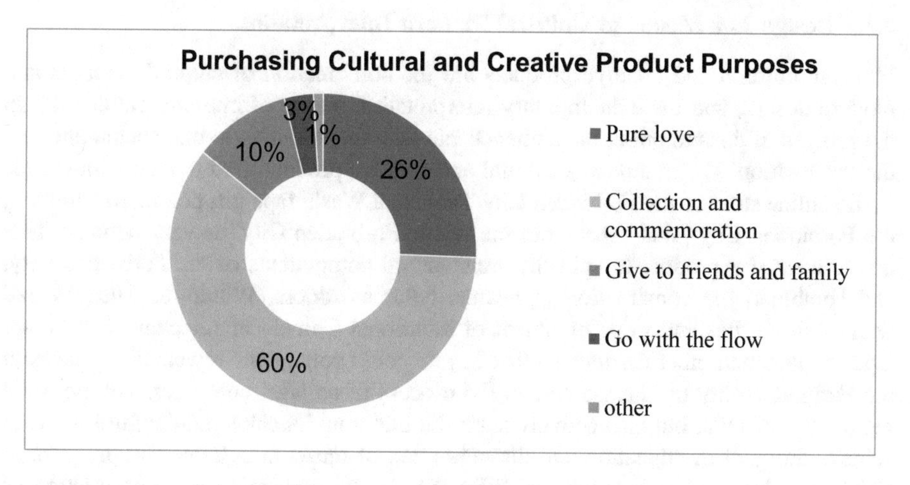

(3) Aesthetic Demand. In the context of consumption upgrading, the simple use value can no longer meet the needs of consumers. In particular, the museum itself is a place where all kinds of artworks, collections and cultural relics are collected. These collections embody the essence of human wisdom from ancient to modern times.

Therefore, consumers will naturally put forward requirements on the artistry of its derivatives, and besides the use value, they also have certain aesthetic value. The development of cultural and creative products in museums should design products that can meet the aesthetic needs based on the production process, materials, shapes, decorations and other aspects of the collections.

(4) Cultural Characteristics of Demand. Each journey is a different experience, so consumers will certainly want to buy different products to commemorate. Cookie-cutter products will make consumers feel tired. Therefore, to enhance the specificity of products and enhance the market competitiveness of products, it is necessary to have product characteristics different from other regions. Nowadays, museums all over the world are actively researching and developing cultural and creative products with their own characteristics and absorbing local cultural characteristics from culture, so as to attract consumers' attention.

Therefore, in order to achieve the purpose of cultural communication of cultural and creative products, it is necessary to subdivide the consumer groups, develop products targeted at different levels such as demand analysis, cultural level, consumption level and aesthetic tendency, and improve them from the feedback of the market. Pay attention to the combination between the design of the product itself and the culture. Different from other products in the market, cultural and creative products are characterized by taking culture as the core. Culture is an essential element in the design process of cultural and creative products. Consumers should be able to feel the cultural connotation in the products and arouse their inner resonance so as to generate cultural identity.

2.2 Design Is a Means to Cultural Modern Interpretation

Museum cultural and creative products are the combination of cultural symbols and modern design, and the contemporary interpretation means of museum culture [1]. In the process of consumption, the audience can take the initiative to understand and feel the connotation. Washi tape is a cultural and creative product used by many museums. In the online store of the Forbidden City Taobao, a Washi tape gift box titled "building the Forbidden City" that "spells out the whole Forbidden City" is very popular. This article created a product to split the architectural components of the Forbidden City, and combined the construction of beams, columns, doors, Windows, corsages, red walls, yellow tiles and ridge ornament of traditional Chinese architecture with Washi tape, stickers and other carriers, so that buyers could collage the towers and palaces in the Forbidden City by themselves. In the process of collage, consumers can not only get the fun of DIY, but also actively learn the traditional architectural culture, so as to achieve the goal of education and dissemination of museum culture. The progress of technology has broken through the difficulties in the replication and preservation of cultural relics, but how to realize the effective dissemination of cultural heritage connotation needs designers' in-depth understanding and innovative interpretation of culture. Figure 1 shows how designers sort out cultural resources and realize the creative transformation process of cultural symbols [2].

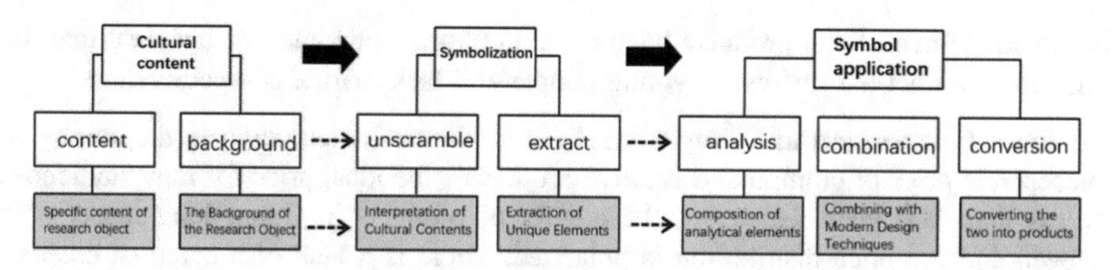

Fig. 1. Cultural and creative product design process.

3 The Practice and Exploration of Cultural and Creative Products in Guangdong Museum

3.1 Practice and Exploration

In 2017, the museum of Guangdong province developed 153 cultural and creative products throughout the year, and the sales volume of cultural and creative products reached 13.3 million yuan, and won four awards including "best cultural product award". In terms of product development, we first tried to cooperate with other companies through joint capital injection, joint research and development and sales sharing. At the same time, the museum strengthens the cooperation with universities through visiting, answering questions, lectures and competitions. In terms of marketing channels, the museum opened Canton Tower counter and other sales channels to actively promote the temporary exhibition of cultural products on commission; Take the initiative to contact the relevant departments of the museum to undertake the exhibition, the museum will be sent to the exhibition hall for sale. In May 2018, Guangdong provincial museum and China international airport business group jointly opened the Guangdong provincial museum experience hall in Guangzhou Baiyun international airport. In addition to putting on the newly developed products of the museum, we can also participate in interactive projects and personally experience the production process of Guang Cai, Guang Embroidery, Duan Inkstone and other excellent Lingnan cultural heritage.

3.2 The Reality of the Dilemma

(1) Lack of Brand Awareness. Nowadays, the competition of cultural derivatives in the market is fierce. If we want to stand out from the competition, we need to form a memory point in consumers' minds, so that consumers can touch their inner feelings when they see the products [3]. However, at present, all kinds of products in the library are operated independently and lack of ties that can be linked together, which makes it difficult to establish a clear brand impression in the minds of viewers.

(2) Products Are Homogeneous and Unattractive. In the survey, the author found that although there are various types of souvenirs sold by Guangdong museum, some of them are not obviously different from the same kind of commodities, and the category homogenization is still a problem, which generally exists in the derivative development of major museums in China. Most of the featured souvenirs sold in Guangdong museum souvenir shops are of primitive and simple shapes, with traditional culture as the main element. Such products have certain cultural connotations, but they may be difficult to attract the interest of young people and lack market competitiveness.

(3) Price Polarization and Single Marketing Channel. According to the survey of "acceptable price of cultural and creative products", the ideal price for most audiences is less than 100 yuan. However, the quality of products in Guangdong museum is uneven and the price distribution is polarized. There is a lack of a batch of creative products with moderate price and cultural characteristics (Table 3).

Table 3. Price survey of acceptable cultural and creative products.

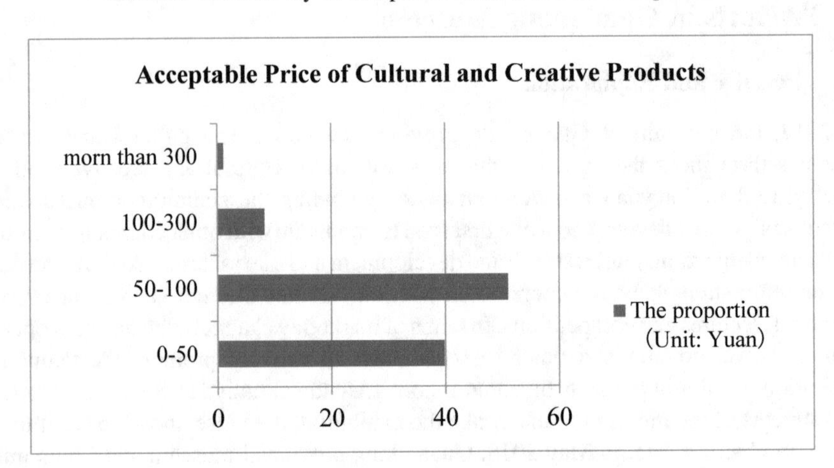

In terms of marketing, apart from the souvenir shop on the first floor, there are three small shops on other floors in the museum of Guangdong province. Meanwhile, counters and experience shops are also set up in Guangzhou tower, Baiyun airport and other places. However, in general, the cultural derivative products of Guangdong museum only support the purchase in physical stores and lack of online purchase channels, which causes inconvenience to consumers when they want to purchase.

4 Cultural Communication Strategies with Cultural Creative Products as the Core

4.1 Shaping Cultural IP to Promote Brand Communication of Museums

In order to achieve sustainable development, museums need public support and recognition, enhance their own differences, facilitate better memory and acceptance of consumers, and attract the public with their own cultural charm. Taking the museum of Xizhou Yandu Site in Beijing as an example, because of its remote location and small scale, its popularity is far less than that of large museums such as the Palace Museum, but they have created the cultural brand of "Talent of Yan" based on their own collection of cultural relics and relics, and designed a series of derivatives and cultural activities based on this, and successfully realized the "cultural relics live, rich". The Museum goes out. As a provincial museum, Guangdong Museum has unique advantages. It is not only huge in scale and geographical location, but also has a deep Lingnan culture as its foundation. These resources can become the cornerstone of shaping its own cultural brand. Guangdong Museums should take Lingnan Culture as a foothold, integrate the resources inside and outside the museum, draw inspiration from the rich collection resources, so as to create a museum brand image with regional characteristics, gain competitive advantages and favorable position, and thus achieve better social benefits.

In addition to shaping the brand image, the museum's rich cultural resources are also a treasure house of IP (intellectual property). The image spokeswoman of Shaanxi History Museum "Tang Niu" is based on the prototype of Tang ladies and maids in the museum and combined with the characteristics of modern cartoons. Once launched, it is very popular. "Tang Niu" is a combination of the profound cultural heritage of the thirteenth ancient capital of Xi'an. Through "Tang Niu", Shaanxi History Museum has held many exhibitions and cultural activities, launched various derivatives such as dolls, pillows, Tuan fans, and produced a series of "Tang Niu chat" cartoons, in order to "Tang Niu" image speaker of China's long-standing history and culture. There have been many cases of IP animation image derived from culture in the world. Kumamoto-ken was originally a little unknown city in Japan. With the support of the local government, designer Hirano designed the mascot image of Kumamon by combining the local culture of the local historic buildings Kumamoto City and the State of Fire. Nowadays, "Kumamon" has become a popular cartoon image of the whole network, which not only improves Kumamoto-ken's popularity unprecedentedly, but also brings huge economic benefits. In today's era of new media, the creation of IP image is not only a mascot, it can become a bridge for the public to understand the museum, make the museum image with thousands of years of cultural heritage younger, close to the masses, close to the masses, and become the source of Museum creative products.

4.2 Using Modern Design for Cultural Transmission

From many successful cases of cultural derivatives, it is not difficult to see that the ingenious combination of cultural connotation and material carrier is an important basis. Culture is the source of inspiration for derivative design [4]. If we just label some handicraft products that look very similar, such products will not be recognized by the public. In the research and development of cultural derivatives, we should keep a good balance between market demand and culture. We should not blindly copy the culture rigidly, nor completely abandon the cultural connotation. Guangdong Museum should rely on its own high-quality culture to design products with Lingnan regional cultural characteristics. Among them, tangible culture includes Lingnan characteristic buildings, treasures of town halls, cultural relics in collection, and intangible characteristic culture such as Cantonese culture, Hakka culture, local language, etc.

Among the cultural derivatives of the Suzhou Museum, there are a series of products, including postcards, bookmarks, stationery and so on, with the museum building as the creative center. Among them, the design idea of stationery pedestal in

"Landscape Interval" comes from the rockery in the landscape garden of Suzhou Museum. This product not only meets the needs of use, but also intuitively shows the regional culture and architectural characteristics of the Suzhou Museum. The architectural noumenon of Guangdong Museum is also a very distinctive building, which integrates traditional culture and modern design. There are also Lingnan region's unique Wok house, Hakka walled house, Chaoshan residence, Lingnan gardens and so on, can become the source of inspiration for product design. At the same time, product design can also take some elements as the core, launch a series of products, make the expression of cultural elements more complete, and enhance the integrity of products.

4.3 To Perfect the Museum and Create a Cultural Transmission System

At present, the construction of cultural and creative industry chain of museums in China is still in the initial stage and has not formed a complete system. To improve this system step by step, on the one hand, we should cultivate talents in the cultural and creative industry, and on the other hand, we should pay attention to expanding product sales channels. First of all, the museum of Guangdong province can cultivate talents in management, design and technology in accordance with its own cultural characteristics according to its actual situation. At the same time, it cooperation with design institutions, colleges and universities, vocational colleges and design organizations to improve the level of product design and development of museums by introducing outstanding college graduates and learning R&D experience from design institutions. It can even gather social forces to encourage the whole people to participate in product development. Such as the "cold room" sticker of the Forbidden City's Taobao, the idea is suggested by the user in the comments area of the microblog. This approach not only broadens product development channels, but also enables consumers to participate in product research and development through interaction, making cultural derivatives more emotionally close to the public.

Due to the public welfare nature of the museum itself, it will be limited in terms of funds, experience and policies. Therefore, it can be considered to cooperate with excellent manufacturers, designers or design companies in the form of authorized development, so as to form complementary advantages with the outside world and facilitate the formation of the cultural and creative industry chain. At the same time to broaden the product sales channels. At present, the sales channels of cultural derivatives of Guangdong museum mainly depend on the souvenir shops in the museum. The online store only has the display function. It is suggested to improve the online store purchasing function of the official website as soon as possible, and realize the parallel of online and offline sales channels. While setting up offline channels such as experience stores and airport stores, we also actively build online sales channels, such as WeChat public platform, small program, Microblog and Taobao exclusive store.

4.4 "Internet+" Opens Up New Ways of History

At present, the Palace Museum, which is in the forefront of the domestic museum culture and innovation industry, has set off a wave by testing the "Internet+", and let people find new ways to open the history [5]. "Forbidden City Taobao" appears on

Microblog, WeChat and other we-media platforms as a very personalized, lovely and humorous image. This completely overturned the previous impression of a serious and solemn, netizens have said "flattered." In order to realize the inheritance and development of culture, the important point of Guangdong museum is to have modern vitality. Look for the intersection of ancient cultural relics and modern life, break the inherent boundaries of the past museum, let cultural relics open the veil of mystery, into our lives. The museum can open the Microblog account and WeChat public account of Guangdong provincial museum, and cooperate with various types of network platforms to actively interact and communicate with netizens. On May 18, 2018, a short film named "cultural relics actor conference" on Tik Tok short video platform was broadcast more than 100 million times in four days, instantly sweeping the WeChat circle of friends. This short video is compiled from the national treasure prototype provided by seven museums including Guangdong provincial museum, which is a successful cross-border cooperation between the museum and video platform.

Apart from the popularity on the Internet, the value of the precious cultural relics handed down for thousands of years in museums should not only exist in tourism. When the emerging industry of the Internet into the traditional culture, it can be injected with new vitality. Guangdong museum should not only explore the activation mode of cultural IP, but also promote the digitization of cultural resources in its collection, and create virtual products derived from culture, such as calendar app with Cantonese cultural style, WeChat small program displaying Duan Inkstone art, and so on. By making use of the Internet's fast and efficient features, Lingnan's unique culture can travel through the distance between time and space, and experience traditional culture without visiting museums. The interlacing of the Internet and museums enables us to see more abundant forms of cultural derivatives of museums, which provides more possibilities for the elaboration of traditional culture.

References

1. Rapaille, C.: The Culture Code. Broadway Books, New York (2006)
2. Xiao, M.: Analysis of cultural creativity and design transformation case. J. Chaoyang **16**, 9–91 (2011)
3. Chang, G.: Demand analysis and transformation in the process of creative product design. J. Decoration **2**, 142–143 (2018)
4. Hong, R., Yang, S., Yang, L., Meng, Y., Hong, Y.: Study on innovative design strategies and approaches of cultural derivatives in shenyang imperial palace. J. Packag. Eng. **38**, 1–6 (2017)
5. Zhang, F.: The development of museum cultural and creative products under the background of "Internet+". J. R. Heritage Prot. **1**, 2–26 (2016)

Bullet Points: Applying Emotional Symbols in Information Management

Amic G. Ho(✉)

Department of Creative Arts, School of Arts & Social Sciences,
The Open University of Hong Kong,
Jockey Club Campus, Ho Man Tin, Kowloon, Hong Kong
amicgh@gmail.com

Abstract. Bullet points are a set of symbols commonly used in typesetting to organise and present different points or notes in a well-organised list. Various shape options for bullet symbols, such as circular, square, dot, or diamond shapes. At the same time, some word processors and publishing design software offer other shapes and colour options. In handwriting, bullet points are usually drawn in any freestyle at the preference of the creator. This could reveal that specific user's select bullet symbols based merely on their preferences without concerning the use of bullet points and icons regarding the nature of communication, not to mention the feeling or mood that these bullets symbols project onto the reader. Misconceptions sometimes create chaos for communication. This study aimed to investigate the nature of the bullet system and how the application of bullet points constructs the information hierarchy. It also assesses the influence of bullet symbols on the organisation and typesetting structure of a document.

Keywords: Emotion · Information management · Typography · Communication design

1 Introduction

In typography design, it is not difficult to find bullet points used to present items in a list. There were various shapes developed as bullet symbols including the circle, square, dot, and diamond. Specific word processing software programs offer more shapes and colour options. In some coding-only text software and some environment where bullet characters are not available some basic symbols, for example, an asterisk (*), hyphen (-), period (.), and lowercase O (o) are conventionally used.

In some cases, the index symbol, which refers to a figure presented as a hand with a pointing finger, was famous for similar purposes: to construct the information hierarchy. In handwriting, bullets can be presented in any style at the preference of the person creating the document. These cases reflect the understanding of some users on the use of bullet points. They seem to mix up the use of bullet points and icons in the process of communication. This misconception sometimes creates chaos when reading and

A. G. Ho (Ed.): AHFE 2019, AISC 974, pp. 32–41, 2020.
https://doi.org/10.1007/978-3-030-20500-3_4

even decreases the credibility of the document. This study aimed to investigate how the application of bullet points constructs the information hierarchy and how their use influences the organisation and typesetting of a document.

2 The Development of the Bullet Point

The bullet point was first developed for use in computer encoding [1, 2]. In past computer systems, bullet points were regarded as glyphs, intended as a set of characters for reading and writing purposes. Glyphs were unique; collective elements used to present specific messages, which were firmly attached to their use from a cultural and social perspective. Another feature of glyphs was their ability to convey some distinction. For example, in Japanese, certain characters are formed using several different marks. In general, these different marks are not classed as glyphs as they do not represent any meaning. Until the addition of other marks, they were differentiated from the distinct characters. In this period, glyphs were not used as characters as certain glyphs could not be displayed on some screens. Until more mature word processors were developed, bullet points were adopted in regular symbols such as o and the asterisk. Some word processors can convert the asterisk (when applied at the beginning of a line) into a bullet point [3, 4]. From this stage, the application of bullet points was developed more practically. The length of the bullet point is not fixed. It can be used to delineate several phrases, a sentence, or a paragraph. In most cases, bulleted items did not end with a full stop, except when the meaning of the points was complete. Certain publications proposed alternative settings: each bulleted line ended with a semicolon, and the last line ended with a full stop. All of these applications were accepted.

3 The Role of the Bullet Point in a Document

Farkas analysed the information presented in the PowerPoint document [5]. He pointed out that documents that applied bullet points were relatively more effective in presenting information. Preparing a presentation without visual supports such as charts, bullet points, etc., would be less effective. According to Frank's analysis, bullet points mediate different pieces of information within a document. He pointed out that bullet points are a useful tool to promote the logical hierarchy in a PowerPoint document if the user edits the ideas carefully. Bullet points provide the framework of a concept and enable the audience to quickly understand the concept.

4 The Categories of Bullet Point

There were different shapes developed as bullet symbols in daily applications [6–9]. Working with words and images: New steps in an old dance. Greenwood Publishing Group Inc. Circular, dot, arrow, and square bullet points were common shapes. In general, word processing applications have hundreds of options in different shapes and colours. Several categories were identified as regular symbols, punctuation, word dividers, general typography, intellectual property, sequence symbols, and graphical dividers (Fig. 1). They included some regular symbols and graphics (Fig. 2). For example, * (named asterisk); - (named hyphen); . (named period); and o (similar shape to lowercase O). These symbols were applied to conventional text-only situations, which were not able to present graphical bullet symbols. There may be more options in handwriting; any style seemed were applicable. In some cases, the index symbol, which in the old style is represented by a hand pointing towards an index finger, was adopted for similar purposes. Lists with bullet points were named as bulleted lists. Sometimes, in internet communication, the element of HTML was applied to structure a list as a non-order list if the list was not structured in numerical sequences.

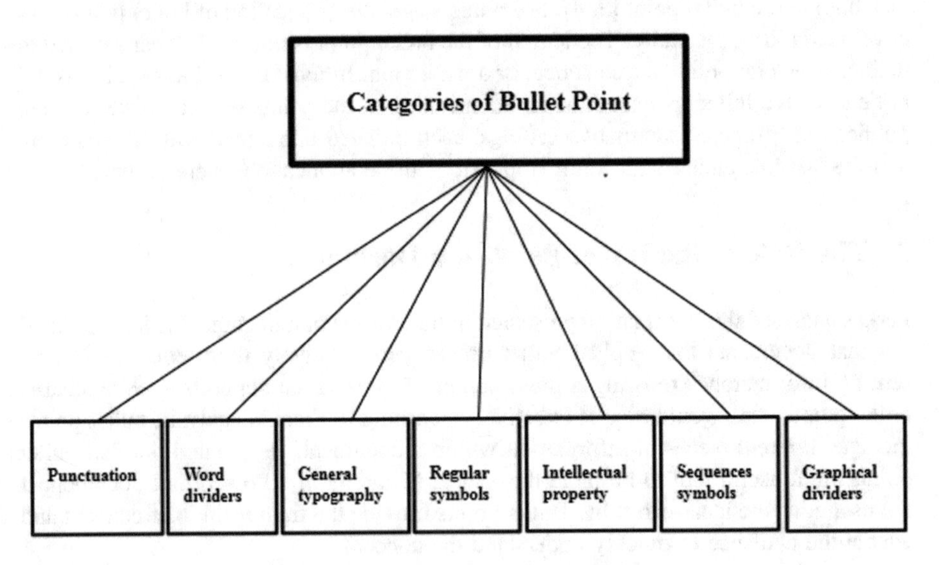

Fig. 1. Sample of a wide figure.

Categoriy name	Bullet symbol name	Shape/form
Regular symbols	Bullet	•
	White bullet	◦
	Triangular bullet	▶
	Bullet operator	∙
Punctuation	Apostrophe	' '
	Brackets	[] () {}
	Colon	:
	Comma	, ، ،
	Dash	– — — —
	Ellipsis	… … ⋯
	Exclamation mark	!
	Full-stop, period	.
	Guillemets	‹ › « »
	Hyphen	‐
	Hyphen-minus	-
	Question mark	?
	Quotation marks	' ' " " ‚ ' „ "
	Semicolon	;
Word dividers	Interpunct	·
	Space	
General typography	Ampersand	&
	Asterisk	*
	At sign	@
	Backslash	\
	Caret	^
Intellectual property	Copyright	©
	Sound-recording copyright	℗
	Registered trademark	®
	Service mark	℠
	Trademark	™
Sequences symbols	Roman numerals	I,II,III
	Alphabetical order	A, B, C
Gaphical dividers	Asterism	⁂
	Fleuron, hedera	❧
	Index, fist	☞
	Interrobang	‽
	Irony punctuation	

Fig. 2. The categories of bullet point and examples.

5 Theories of Information Hierarchy Construction

Hayama, Nanba, and Kunifuji found various electronic documents, and they emphasised that structuring information in visual ways was important for a human to process the information from the data [6]. They conducted a study to investigate how to structure data to be information those were ease for understanding effectively. They proposed that the first steps were to clarify the relationship between the data and then organise the data as units. For example, the title, the body of text, keywords, images, etc. Based on this organisation, the hierarchy could be built from top to bottom.

At the same time, Allen pointed out that the process of editing influences the audience's understanding of the content. Certain scholars mapped out several possible elements for constructing the information hierarchy [10–12]. They proposed that it was necessary to understand how the nature of messages changed from the point of receipt to motivating further action. It was found that for a given message, no matter what the content related to, data were the primary stage those audiences obtained. Data were facts, both non-organised and analysed. Data were conceived as symbols that represented stimuli. Certain scholars commented that data were not yet usable. Sharma proposed that data should be organised and compared to obtain the real meaning of the message [13]. In some situations, data were regarded as more than just symbols; it included signals which may be subjective and depend on observations. In other words, these signals included certain emotions and were processed as information. Information was a sensory stimulus in empirical perception. Also, it contained descriptions, and it was identified as useful from the group of data. It was expected to provide specific answers to interrogative probing such as who, what, where, how many, and when.

6 The Function of Bullet Points in the Information Construction Process

Considering the above investigation, bullet points are useful as a mediator between different pieces of information within a document [8, 10] (Fig. 3). Bullet points construct the logical hierarchy of the document through their graphic presentation. This graphic presentation stimulates different emotions within an audience. Emotions influence the approaches taken by the audience to process information and understating on information hierarchy those leading audience quickly understand the features of the concept. Information was identified from the content of data under the influence of the audience's changed emotions Information was know-what. Referring to such processing of information, the nature and content of the information are judged. Knowledge is developed by making explicit in symbolic form and generated based on the obtained information. In other words, the application of bullet points influences the audience's emotions and judgments about the topic.

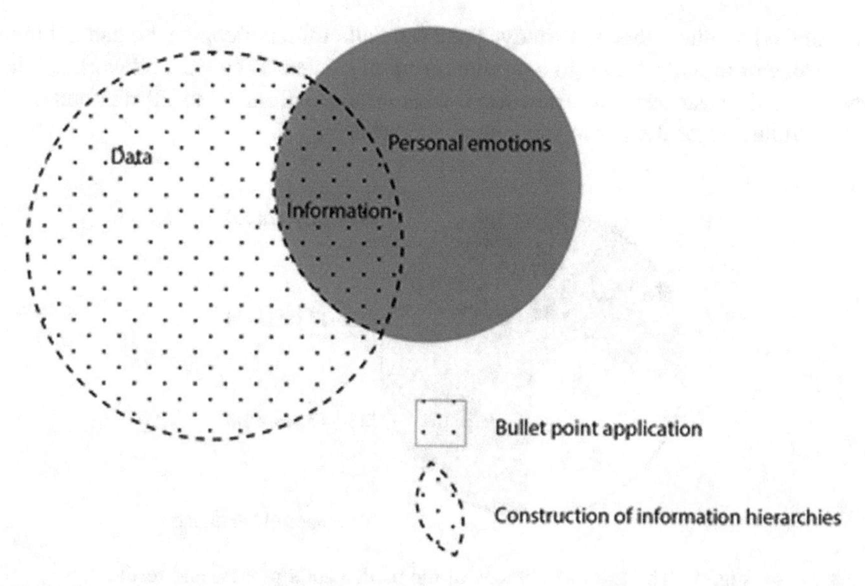

Fig. 3. The relationship between data, information, personal emotions, and construction of the information hierarchy.

7 The Application of Bullet Points Influences the Audience's Emotions

Following the proposed functions of bullet points in the information construction process, it was found that the audience's emotions and judgments are affected by the organisation of bullet points and the construction of the information hierarchy. Also, Frank [5] pointed out that bullet points in PowerPoint mostly applied graphical bullet symbols because these symbols catch the audience within the presentation. Bullet points deliver specific messages dependent on the selected bullet symbols. Information is relatively subjective, and its meaning can be influenced by the attached symbols. The state of awareness of information is consciousness. It was the physical manifestation form. The information represents both the emotional and affective state. It was found that physical counterpart included part of the emotion or affective state. Moreover, therefore, one of the differences between data and information is the part of cognition contained in the organised information.

8 Pilot Study for Understanding Knowledge of Ordinary People About Bullet Point

In order to understand how ordinary people know about the categories and function of bullet point, a survey was conducted. The objective of the lesson was for students to consolidate the conceptual understanding of bullet point and examine their function in the process of information hierarchy construction. One hundred twenty participants were invited to attend this pilot study. They were all ordinary people who had not much knowledge or training in design or communication studies. They were all working class who worked in various industries. Their ages arranged from 18 to 50 randomly. The age distributions of the participants were shown in Fig. 4.

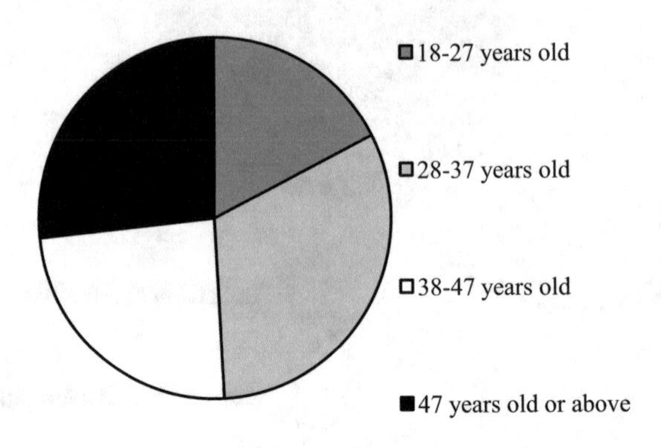

Fig. 4. The age distributions of the participant's process and result.

In order to consolidate their conceptual understanding of bullet point, participants in this pilot study were asked to recognise the categories of bullet point and compare their functions in their daily information processing tasks. Some feedback from participants based on their experience of applying bullet point was shown below (Table 1). 70% of participants recognise the different categories of bullet point. All kinds of bullet point could deliver messages for communication. The participants selected punctuation, graphical dividers and sequence symbols as the most effective kinds of bullet point for communication. All participants agreed that the bullet point could evoke their emotional concerns.

Table 1. The table presented feedback from participants about their roles in the pilot study.

	Survey (Tick the appropriate box, rate from 1 to 5, 1 = Strongly Disagree, 5 = Strongly Agree)				
Regarding the provided options of the bullet point	1	2	3	4	5
I would recognize the different categories of bullet point	5%	25%	50%	20%	–
I think the 'punctuation' is the most applicable for communication	–	–	10%	40%	45%
I think the 'word dividers' is the most applicable for communication	5%	–	50%	40%	5%
I think the 'general typography' is the most applicable for communication	–	–	–	80%	20%

(*continued*)

Table 1. (*continued*)

	Survey (Tick the appropriate box, rate from 1 to 5, 1 = Strongly Disagree, 5 = Strongly Agree)				
Regarding the provided options of the bullet point	1	2	3	4	5
I think the 'intellectual property' is the most applicable for communication	5%	–	50%	40%	5%
I think the 'sequence symbols' is the most applicable for communication	–	–	–	80%	20%
I think the 'graphical dividers' is the most applicable for communication	–	–	–	25%	75%
I think the 'general typography' can help me to express the emotional messages most effectively	–	–	–	80%	20%
I think the 'word dividers' can help me to express the emotional messages most effectively	5%	–	50%	40%	5%
I think the 'intellectual property' can help me to express the emotional messages most effectively	–	–	–	80%	20%
I think the 'sequence symbols' can help me to express the emotional messages most effectively	–	–	–	25%	75%
I think the 'graphical dividers' can help me to express the emotional messages most effectively	–	5%	25%	45%	25%
I think the bullet point would evoke my emotional concerns	–	–	–	25%	75%

9 Conclusion

This study investigated the nature of the bullet system. Bullet points are a set of symbols commonly used in typesetting to organise and present different points or notes in a well-organised list. Nowadays, there are various shape options for bullet symbols. Most commonly used are the circular, square, dot, or diamond shapes. At the same time, some word processors and publishing design software offer other shapes and colour options. Certain index symbols, for example, a pointing finger, a tick, or even a small pencil are also famous for similar purposes: to construct the information hierarchy. In handwriting, bullet points are usually drawn in any freestyle at the preference of the creator. This could reveal that specific user's select bullet symbols based merely on their preferences without concerning the use of bullet points and icons regarding the nature of communication, not to mention the feeling or mood that these bullets symbols project onto the reader. The misconceptions sometimes create chaos for reading and even decreases the credibility of the document. This study aimed to investigate the nature of the bullet system and how the application of bullet points constructs the information hierarchy. It also assesses the influence of bullet symbols on the organisation and typesetting structure of a document.

To conclude, in communication design, bullet points are intended to organise and present different points or notes in a list by use of a set of symbols. Various shapes are applied depending on the document and its intended purpose. Several categories of symbol were identified including regular symbols, punctuation, word dividers, general typography, intellectual property, sequence symbols, and graphical dividers. It was identified that the development of word processing and publishing design software strives to offer more shapes and colour options. Theoretical findings revealed that users select the bullet symbols based on their personal experience and knowledge, and at the same time, they must consider the use of bullet points and icons regarding the nature of communication. It would be more effective if they were to consider the feeling or mood that these bullets symbols project onto the reader. Well-organised bullet-pointed lists would ease reading and increase the credibility of the document. The application of bullet points and their function on constructing the information hierarchy, as well as their capacity to affect the organisation and typesetting structure of a document were understood.

Acknowledgements. The author would like to say thank you to the research participants for their help and involvement.

References

1. Bly, R.W.: The case against PowerPoint. In: Successful meetings, vol. 50, no. 12, pp. 50–52. Proquest ABI/INFORM (2001)
2. Blokzijl, W., Andeweg, B.: The effects of text slide format and presentational quality on learning in college lectures. In: Proceedings of International Professional Communication Conference, pp. 288–299 (2005)
3. Atkinson, C.: Beyond Bullet Points: Using Microsoft PowerPoint to Create Presentations that Inform, Motivate, and Inspire. Pearson Education, London (2011)
4. Clair, K.: A Typographic Workbook: A Primer to History, Techniques, and Artistry. Wiley, Hoboken (1999)
5. Farkas, D.K.: Managing three mediation effects that influence PowerPoint deck authoring. Tech. Commun. **56**(1), 28–38 (2009)
6. Hayama, T., Nanba, H., Kunifuji, S: Structure extraction from presentation slide information. In: Pacific Rim International Conference on Artificial Intelligence, pp. 678–687. Springer, Berlin (2008)
7. Farkas, D.K.: Toward a better understanding of PowerPoint deck design. Inf. Des. J. **14**(2), 162–171 (2006)
8. Allen, N.: Working with Words and Images: New Steps in an Old Dance. Greenwood Publishing Group Inc., Westport (2002)
9. Condon, W.: Selecting computer software for writing instruction: some considerations. Comput. Compos. **10**(1), 53–56 (1992)
10. Maeda, K., Hayashi, Y., Kojiri, T., Watanabe, T.: Skill-up support for slide composition through discussion. KES **3**, 637–646 (2011)
11. Bahrani, B., Kan, M.Y.: Multimodal alignment of scholarly documents and their presentations. In: Proceedings of the 13th ACM/IEEE-CS Joint Conference on Digital Libraries, pp. 281–284. ACM (2013)

12. Tanaka, S., Tezuka, T., Aoyama, A., Kimura, F., Maeda, A.: Slide retrieval technique using features of figures. In International MultiConference of Engineers and Computer Scientists, vol. 1, pp. 424–429 (2013)
13. Balog, K.P., Gašo, G.: User satisfaction survey in the Faculty of Humanities and Social Sciences Library–do we manage to deliver? J. Res. Writ. Books Cult. Heritage Inst. **9**(1) (2016). Libellarium
14. Sharma, N.: The origin of the data information knowledge wisdom (DIKW) hierarchy (2008)
15. Bellinger, G., Castro, D., Mills, A.: Data, Information, Knowledge, and Wisdom (2004)
16. Boiko, B.: Content Management Bible, 2nd edn, p. 57. Wiley, Indianapolis (2005)
17. Bosancic, B.: Information in the knowledge acquisition process. J. Doc. **72**(5), 930–960 (2016)
18. Gamble, P.R., Blackwell, J.: Knowledge Management: A State of the Art Guide, p. 43. Kogan Page, London (2002)

Impact of Inspiration Sources on Designer's Idea Generation Strategy

Xinyu Yang and Jianxin Cheng(✉)

School of Art, Design and Media, East China University of Science and Technology, No. 130, Meilong Road, Xuhui District, Shanghai 200237, China
546467089@qq.com, 13901633292@163.com

Abstract. The idea generation phase in the design process is often referred to as 'the fuzzy front end', which combing inspiration from a large amount of stimuli and pilot experience into concepts. How to lead designers is a significant task in design method research. This paper is firstly aim at visualising the idea generation and concept deepening processes of designers with different levels of design experiences. Second, to find out the influence of inspiration sources with different stimulation distance on designers' creativity, and classify designers' preference of inspirational stimuli. This paper select design students and expert designers who have more than two years experience as test subjects. Two groups of subjects are asked to conduct same design task within a certain period of time. The design task involves three sub-sessions, each of which provides a type of stimulus information, including text, pictures and design plans, representing different levels of stimulation distance, and requires the subjects generate design alternatives as much as possible based on the given stimuli. Observe the subjects' design outputs under different inspiration sources, using the extended linkography as a tool to visualise designers' problem space evolution. It measures the designers' creativity and fixation preferences of different inspiration sources by analysing the number of design alternatives, the number of transformation links and stimuli distance. This paper may have significance for the study of idea generation methods and the development of related tools.

Keywords: Idea generation · Stimuli · Extended linkography · Inspiration sources

1 Introduction

The idea generation phase in the design process is often referred to as 'the fuzzy front end', which combing inspiration from a large amount of stimuli and pilot experience into concepts. How to lead designers is a significant task in design method research.

Earlier research has in-depth exploration on the impact of textual descriptions and visual information [1] as stimuli on designers' idea generation. However, those studies ignore important details of the real design process and individualised differences of designers in order to emphasise the impact of certain stimulus on inspiration. Based on the previous researches, this paper firstly aims to use visual diagrams to sort out the concept divergence and deepening process of designers with different experiences.

A. G. Ho (Ed.): AHFE 2019, AISC 974, pp. 42–52, 2020.
https://doi.org/10.1007/978-3-030-20500-3_5

Secondly, we will find out the influence of different distance of inspiration to designers' creativity, thus classifying designers' stimulating preference. In order to analyse the process of idea generation more effectively, this paper uses the 'extended linkography' proposed by Cai et al. [2] to describe experiment process, and analyses participants' inspirational time period and reasons through interviews and 'link density calculation [3]. This study tests the influence of different stimuli on design behaviour, and has certain meaning for design method research and idea generation tools.

2 Former Research Findings

2.1 Stimuli and Idea Generation Strategy

Several studies from different theoretical backgrounds have investigated the role of stimuli as sources of inspiration. Association theory believes that the creation of new ideas is the result of integration, which can be promoted through unique stimuli [3]. When designers receive external stimuli as inspiration sources, stimuli form search clues in short-term memory to explore long-term memory to simplify knowledge acquisition and inspire creativity [4]. Another important strategy is analogical reasoning, which refers to the ability to perceive and use similarities between relationships, and is widely recognised as a basis for scientific and artistic creativity [5–7]. Analogy is also a basic cognitive process, such as learning, forecasting, reasoning and problem solving. Goldschmidt defined analogy as a process of mapping and transfer from one situation to another based on 'the similarities between relationships'. The process of analogical mapping is a two-way operation: from a known example to abstraction, and from abstraction to a new candidate-example to solve the problem in hand. Visual analogy has been shown to be an important strategy in solving ill-defined problems for both novices and experienced designers [8].

Association and analogy could map and transfer information through stimuli in forms of image, sounds, words, etc. Malaga conducted an experiment to observe difference of participants' creation by showing them words, pictures and combination of words and pictures as stimuli [9]. He mentioned in his paper that pictures is more possible to inspire creativity than others, but the impact of stimuli may depend on the person who perceives them. Also, Yuan and Song [3] claims that depend on the distance of stimuli, designers' creativity level can be classified into 'long distance active' and 'close distance active'. 'Long distance active' designers are good at abstract intangible elements from stimuli to generate alternatives, while 'close distance active' designers tend to look for stimuli with tangible design knowledge to inspire ideas. In addition, designer's creativity level is also affected by other factors, for instance, experienced designers are more likely to abstract inspiration from their own experience, and reflect on those ideas to create more alternatives. In contrast, design novices can create more imaginative ideas because they are less subject to technical requirements.

In sum, design is a complex, contradictory and non-linear process. Idea generation is at the early stage of the whole design process. Its formation is inseparable from the acquisition of various external sources, an it also shows the characteristics of cognition and processing of human brain. Different forms of inspiration can have different effects on design performance, and these effects vary from level of professional.

2.2 Problem Space

The design thinking of idea generation process is reflected in two dimensions of problem space development, namely the depth and breadth of thinking [10]. The breadth of thinking represents the horizontal transformation in inspiration process, which usually means the diversification of ideas. While the depth of thinking represents the vertical development in inspiration process, and it reflects the completeness of the concept. Therefore an active design conception should both develop in depth and breadth in the problem space.

The difference in designer thinking is reflected in the development mode of its problem space. If we want to discuss the problem space mode of designer's idea generation, we should break down the inspiration process to build analysis model and understand the dynamics of its design performance.

2.3 Linkography

Goldschmidt proposed a more fine-grained assessment of design performance and linkage between stimuli and concepts named 'linkography' in 1990 [11]. Linkography creates a connection for all design moves in the idea generation phase through coding and visualisation. It contains the following basic elements (Fig. 1): first, a design move, which can be a step, an act or an operation. Van der Lugt believes that design moves need to have clear relationship with solutions of the current task, and have signs of concept delivery [12]. Second, a link (or a node) that used to determine the relationship between design moves. Related design moves can generate a link, which is the node at the intersection of diagonal network lines connecting two related moves. Finally is the critical move containing a lot of forelinks (links that connect with former moves) and backlinks (links that connect with later moves). Higher number of forelinks indicate that the critical move incorporates many previous concepts, while higher number of backlinks indicate that the move inspires many subsequent concepts.

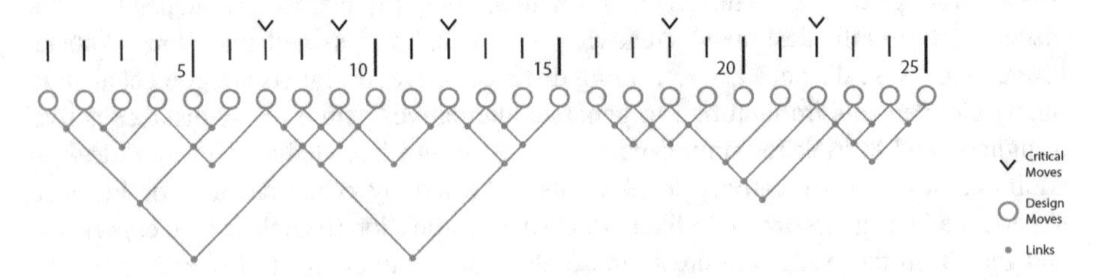

Fig. 1. Linkograph of design cycle [11]

Goldschmidt's link coding method relies on common sense, and is judged by the principle that 'the two design moves are from the same or similar theme, the link is established', which have certain ambiguity. Therefore, many scholars proposed 'link type index' (LTI) as measurement of correlation of design moves. Van der Lugt [12] extended links into three types such as 'supplementary' (S), 'modification' (M) and 'tangential' (T) links. The 'supplementary' link refers to small and auxiliary change under the same concept; the 'modification' link indicates structural changes; the 'tangential' links means a link that is not directly related to previous concepts. Perttula and Sipila classified links as 'parts sharing' (PS), 'same principle' (SP) and 'modification' (MD), and assigned different weights to develop a metric named 'weighted link density' [13]. Higher 'link density index' represents a higher genealogical linkage between design object and solutions, which is a symbol of design fixation. Linkography is also applied to other areas and developed different forms. Several key adaptations of linkography is shown in Table 1 [14–18].

Table 1. History of linkography

Author	Linkography development
Goldschmidt [11]	Proposed idea of 'linkography'
Van der Lugt [12]	Proposed 'link type index' (LTI), extended links into three types such as 'supplementary' (S), 'modification' (M) and 'tangential' (T) links
Vidal et al. [14]	Created a more condensed linkography by grouping discrete ideas into 'global ideas'
Kan and Gero [15]	Followed Shannon's information theory to calculate entropy (unpredictability) in a linkograph, solving the problem that the simple link density calculation may favour overly saturated linkographs
Perttula and Sipila [13]	Created 'weighted link density', and assigned weights for four design move types such as 'new concept', 'part sharing', 'same principle' and 'modification'
Kan and Gero [15]	Cluster analysis of link distance and density along X and Y axis
Kan and Gero [16]	Coded design moves using function, behaviour and structure ontology, to show the distribution of solution focus throughout the ideation process
Cai, Do and Zimring [2]	Added vertical links to linkography to represent the depth of idea generation
Pourmohamadi and Gero [17]	Developed LINKODER software
Cash and Stroga [18]	Developed link networks to understand the ideation process more in details
Jiang and Gero (2017)	Focus on developing tools for linkography of conversational turns rather than individual design moves
Yuan and Song [3]	Classified designers' stimuli preference into four types such as 'long distance active', 'close distance active', 'overall active' and 'overall inactive'

Goldschmidt found that the distribution of links among design moves could represent the general pattern of design process [12]. She proposed three patterns of linkograph: 'chunk', 'web' and 'sawtooth track'. A chunk is a group of moves that are almost exclusively linked among themselves, it shows cycle of thought and investigation of sub-sessions. A web is a large number of links among a relatively small number of moves, it refers to a period of intense idea generation. And sawtooth track is a certain sequence of linked moves, which to some extent represents generation of inspiration, though comparing to 'chunk' and 'web', it is insufficient on depth and breadth of problem space development.

Cai and his colleagues proposed 'extended linkography' (Fig. 2) based on Goldschmidt's linkography [2]. It keeps the key elements of linkography: design move and link, adding 'lateral transformation' (LT) and 'vertical transformation' (VT) to the original linkography. A 'lateral transformation' is from one idea to a slightly different idea or an alternative idea. In a 'vertical transformation' the movement is from one idea to a more detailed idea. In other words, lateral movements may be relevance or irrelevance with each other, while vertical movements must have relationships. Extended linkography have better reflection on designer's idea generation line, which helps provide more insight about the impact of inspiration source on design performance.

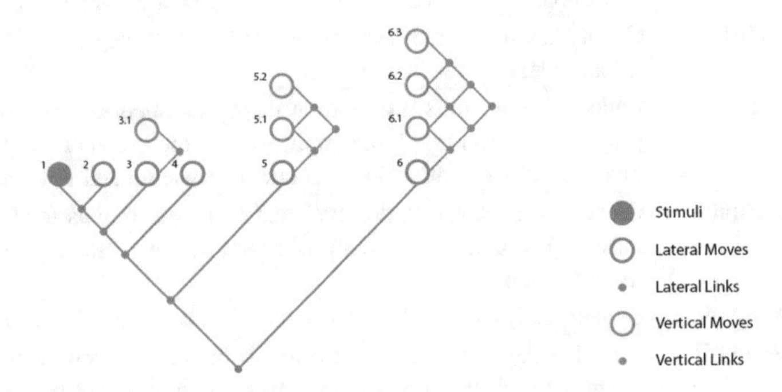

Fig. 2. Extended linkography [2]

3 The Experiment

3.1 Experiment Overview

This research conducted a pilot experiment to investigate to what extent the three types of stimuli as inspiration sources affect the performance of design experts and novices. Three specific stimuli are: (1) textual description; (2) abstract images; (3) plans of cars in sale. The design task had three sub-sessions, participants were asked to generate as many design solutions for cars as they could with inspiration source given prior to design process. To avoid the influence from the previous tasks to the subsequent tasks, each sub-session were set to a different object. Session 1 was to create design schemes of sports car with the stimulus of textual description of a car. Session 2 was to generate design schemes of electric vehicles with the stimulus of abstract images (no direct relationship with cars). Session 3 was to produce business car solutions with the stimulus of pictures of well-known high-end car schemes on the market. Each sub-session time is 30 min, after that was interviews to each participants, explaining their feelings and ideas. To simplify the task, participants were asked to only sketch the front face of the car. Participants could not use removable pens to sketch in order to obtain the complete ideation process. In addition, participants could conduct the task at their daily working place alone.

3.2 Participants

This experiment invited three undergraduate students from East China University of Science and Technology, majored in industrial design, and two professional car designer with more than three years experience. The design expert group (subject A and B) has a wealth of design working experience, during the ideation process is more likely to use their own knowledge and various forms of stimuli to generate solutions. The novice group (subject C, D and E) has established relatively complete design knowledge system through more than three years' technical learning and training. But because of their lack of experience, design novices may rely more on external stimuli.

3.3 Coding Scheme

The design alternatives generated in the experiment were coded based on the coding method in the existing research [2]. Participants were numbered as A, B, C etc. Three sub-sessions were coded according to the English initials of the stimuli provided: textual stimulus as 'T'; image stimulus as 'I'; plan stimulus as 'P'. The concepts drawn by the subjects were sorted by Arabic numerals, and if one concept contained several sketches, called process sketch, would be marked as 1.1, 1.2 etc. Thus the whole number, for example: B-T-2.2 refers to the second process sketch of the second design alternatives generated by subject B in text sub-session.

The classification of link type was adjusted based on the LTI established by van der Lugt [12] and Perttula et al. [13]. According to the distance between concepts, the new link type was sorted from close to far as four stages: 'concept deepen', 'tangible element sharing', 'intangible element sharing' and 'new concept'. The consideration of 'concept deepen' was equal to vertical transformation link (LVT); 'tangible/intangible element sharing' referred to lateral transformation link (LLT); 'new concept' did not contain links. The division helped researchers to determine the relationships between concepts and to make linkograph more reasonable. The sketches and extended linkograph of subject E in text sub-session could be seen in Fig. 3 as an example.

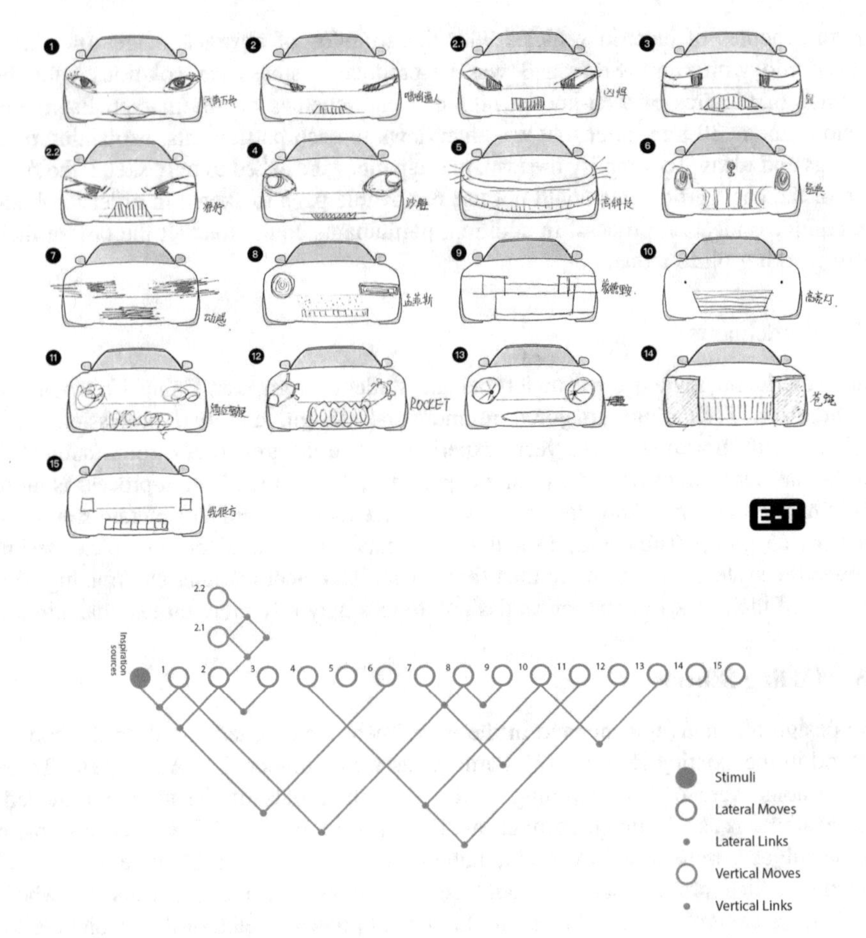

Fig. 3. Sketches and extended linkograph of subject E in text sub-session

4 Result and Analysis

The results showed different responses of five subjects to various stimuli, and individual differences of creativity. In this section, we firstly summarised and qualitatively analysed each one's design performance, and then visualised and evaluated the experiment results through 'extended linkograph'.

This experiment has collected fifteen samples from five subjects, and generated fifteen groups of linkograph. In order to cross-compare the design performance related to different inspiration sources and designers' experience level, researchers arranged all graphs together into 3-by-5 matrix. Each row represented one subject, and each column represented on inspiration source. Figure 4 shows the aligned extended linkograph.

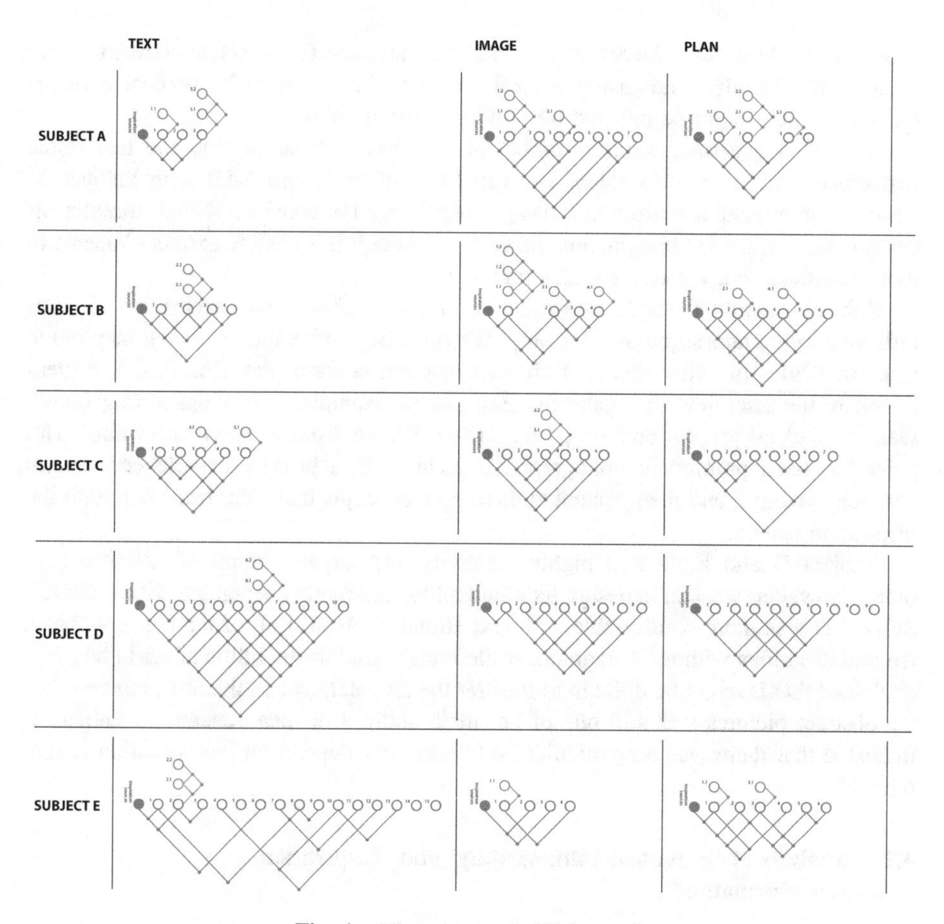

Fig. 4. Aligned extended linkograph

4.1 Qualitative Analysis

Overall, subject A and B, as design experts, produced similar numbers of sketches in the experiment, and had better capabilities in concept deepening rather than novices. Their stimuli preference was more inclined to abstract visual stimulation. While subject C, D and E, as students, had better performance in the breadth of idea generation than experts. Subject D and E preferred text stimulation, while subject C had no obvious stimuli preference.

Subject A generated less design alternatives in tex sub-session than in the other two sub-sessions, and even finished exhausted ideas before he reached the time limit. He mentioned in the later interview: 'It is hard to imagine a matching look from the text, but the picture gives me more inspiration'. However, in the extended linkograph, subject A could make a deepening action almost on average in each sub-session. That means after laterally analogising several concepts, he would make vertical transformations, which might be affected by previous design habits.

Subject B generated similar number of sketches with subject A, and had stable performance in each sub-session. He also has similar design habit with Subject A, which is divergent and deepen concepts regularly. He considered that stimulus of design plan limited his imagination, he said: 'Although it is easy to extract elements in those solutions, but the result is like imitated.'

Subject showed a stable performance in each sub-session, having no obvious influence by certain stimulus. Her outputs kept a high relevance with the inspiration sources, while she felt difficult to abstract inspiration form plan stimulus. She mentioned in the interview that 'take the Benz as an example, it contains strong family features, and all proportions are perfect. Once I learn from it, it will be weird'. Her point has been proofed in linkograph C-P. Her design process experienced several 'surface analogy', and then generated three new concepts that were irrelevant with the inspiration source.

Subject D and E showed higher creativity and larger amount of sketches than others, however, weaken in design fixation ability, and rarely deepen the critical moves. Subject D was more comfortable with text stimulus. Almost all of the concepts came from associations with the paragraph, while image stimulus had little effect to her. She explained that it might be difficult to transfer the line elements in the image to cars, but the abstract pictures was still one of her main sources of idea generation. Subject E indicated that there was no particular preference, but depend on her condition at that time.

4.2 Analysis of 'Extended Linkography' and 'Inspiration Transformation'

In order to measure levels of creativity of the subject in each sub-session, this research uses the 'link density index' (LDI) [13], 'lateral link density index' (LDILT) and 'vertical link density index' (LDIVT) [2] to describe the micro level information of the design process. LDI is the ratio between the number of total links and the number of total design moves. LDILT and LDIVT added the concepts of 'lateral transformation' and 'vertical transformation'. LDILT is the ratio between the number of lateral transformation links and the number of total design moves. LDIVT is the ratio between the number of vertical transformation links and the number of total design moves. Besides, this study also focuses on the impact of stimuli on design performance. Thus the concept of 'inspiration transformation' is considered, which represents the concepts transferred from the inspiration sources. Links that connecting inspiration source and design moves are considered as 'inspiration transformation links'. LDIIT is the ratio between inspiration transformation links and total design moves. In other words, the larger the value of LDIIT, it is easier for the subject to extract inspiration from the stimuli. Based on the above information, the link density calculated for the samples of this experiment is shown in Table 2.

Table 2. LDI, LDIIT, LDILT and LDIVT of different subjects in each sub-session

	LDI	LDI_{IT}	LDI_{LT}	LDI_{VT}	LDI	LDI_{IT}	LDI_{LT}	LDI_{VT}	LDI	LDI_{IT}	LDI_{LT}	LDI_{VT}
	Subject A				Subject B				Subject C			
Text sub-session	1.14	0.86	0.57	0.57	1.43	0.71	1	0.43	1.63	0.75	1.38	0.25
Image sub-session	1.55	0.91	1.18	0.36	1.60	0.80	0.80	0.80	1.56	0.67	1.22	0.33
Programme sub-session	1.33	0.89	0.78	0.56	1.78	0.89	1.56	0.22	1.25	0.50	1.25	0
	Subject D				Subject E							
Text sub-session	2.92	0.85	2.69	0.23	0.78	0.28	0.61	0.17				
Image sub-session	1.33	0.11	1.33	0	0.67	0.50	0.50	0.17				
Programme sub-session	0.90	0.50	0.90	0	1.44	0.67	1.22	0.22				

First, comparing the expert group with the novice group, we can find that the expert group shows higher LDI than the novice group, which seems to be an indicator of higher creativity. But as we include LDILT, we find novice group has better performance on breadth development of the problem space. On the other hand, the LDIVT of novice group is lower than expert group's, which demonstrate the weakness of novice group on concept refinement and previous moves review. Second, although some of the subjects mentioned the difficulty of inspire from plan stimulus, but comparing the value of LDIIT of plan session to the others, plan stimulus shows higher inspiration extraction than the other two stimuli. Moreover the sketches of plan session show good creativity as well, which probably because the plan stimulus has shorter distance to the design aim, and it is easier to analogise, while it is also hard to break through the existing framework.

5 Conclusion and Future Work

Designers will use different strategies when receiving various stimuli. Therefore, for the development of inspiration tools, it is necessary to consider both the suitability of inspiration source and design object, and designers' preference of stimulus distance. This research tested the influence of different stimuli as inspiration sources on design performance, and impact of design experience level on restricted design activities. We used LLT index and LVT index to demonstrate the movement from inspiration source to design moves, which helps explore more details in the design behaviour of human being.

The findings cannot be generalised due to the small sample size, for instance, the possibility of universal similarities of design experts' idea generation process. In addition, this study does not consider the impact of the order of stimuli exposure as we need to simplify the experiment process. All the stimuli in this experiment are presented in the same order, however, for experienced designers, the experience in an earlier sub-session may create influence for the latter sub-sessions. In future research, we will conduct research by increasing the number of participants and disrupting the order of stimuli exposure.

Acknowledgement. This paper was supported by 'Shanghai Summit Discipline in Design' under Grant No. DC17013 (master studio project of Regional Characteristic Product Research and Development for 'The Belt and Road Initiatives').

References

1. Goldschmidt, G., Smolkov, M.: Variances in the impact of visual stimuli on design problems solving performance. J. Design Studies. **27**(5), 549–569 (2006)
2. Cai, H., Do, E., Zimring, C.: Extended linkography and distance graph in design evaluation: an empirical study of the dual effects of inspiration sources in creative design. J. Design Studies. **31**(2), 146–168 (2010)
3. Yuan, X., Song, D.: Impact of pictorial stimuli on designer's concept generation strategy. J. Packaging Eng. **39**(14), 166–171 (2018)
4. Mednick, S.: The associative basis of the creative process. J. Psychol. Rev. **69**(3), 220–232 (1962)
5. Nijstad, B., Stroebe, W., Lodewijkx, H.: Cognitive stimulation and interference in groups: exposure effects in an idea generation task. J. Exp. Soc. Psychol. **38**(6), 535–544 (2002)
6. Han, J., Shi, F., Chen, L., Childs, P.: A computational tool for creative idea generation based on analogical reasoning and ontology. J. Artif. Intell. Eng. Des. Anal. Manuf. AIEDAM **32**(4), 462–477 (2018)
7. Goldschmidt, G.: Visual analogy: a strategy for design reasoning and learning. In: Design Knowing and Learning: Cognition in Design Education. Elsevier, Amsterdam (2001)
8. Casakin, H., Goldschmidt, G.: Expertise and the use of visual analogy: implications for design education. J. Des. Stud. **20**(2), 153–175 (1999)
9. Malaga, R.: The effect of stimulus modes and associative distance in individual creativity support systems. J. Decis. Support Syst. **29**(2), 125–141 (2000)
10. Kudrowitz, B., Wallace, D.: Assessing the quality of ideas from prolific, early-stage product ideation. J. Eng. Des. **24**(2), 120–139 (2013)
11. Goldschmidt, G., Tasta, D.: How good are good ideas? Correlates of design creativity. J. Des. Stud. **26**(6), 593–611 (2005)
12. Van der Lugt, R.: Developing a graphic tool for creative problem solving in design groups. J. Des. Stud. **21**(5), 505–522 (2000)
13. Perttula, M., Sipila, P.: The idea exposure paradigm in design idea generation. J. Eng. Des. **18**(1), 93–102 (2007)
14. Vidal, R., Mulet, E., Gomez-Senent, E.: Effectiveness of the means of expression in creative problem-solving in design groups. J. Eng. Des. **15**, 285–298 (2004)
15. Kan, J., Gero, J.: Acquiring information from linkography in protocol studies of designing. J. Des. Stud. **29**(4), 315–337 (2008)
16. Kan, J., Gero, J.: Characterizing innovative processes in design spaces through measuring the information entropy of empirical data from protocol studies. J. Artif. Intell. Eng. Des. Anal. Manuf. AIEDAM **32**(1), 32–43 (2018)
17. Pourmohamadi, M., Gero, J.: LINKOgrapher: an analysis tool to study design protocols based on FBS coding scheme. J. Int. Conf. Eng. Des. **2**, 294–303 (2011)
18. Cash, P., Storga, M.: Multifaceted assessment of ideation: using networks to link ideation and design activity. J. Eng. Des. **26**(10–12), 391–415 (2015)

Design in Advertising and Media Communication

Emotions in Advertising: How Emotions Affect Creativity and Communication

Amic G. Ho(✉) and Sunny Sui-kwong Lam

Department of Creative Arts, School of Arts and Social Sciences, The Open University of Hong Kong, Ho Man Tin, Hong Kong
amicgh@gmail.com, ssklam@ouhk.edu.hk

Abstract. Advertising delivers messages in innovative ways to attract target audience. While some evidence shows that emotions, as well as cultural values, are key factors affecting the process of advertising and communication. In this paper, the influence of human factors, including social recognition and emotion concerns in the communication design process. However, few studies investigated how human factors can enhance creativity and interact with the communication process. Hence, this research study aimed at investigating the relationships between emotion and social recognition to motivate the interactions between the roles in the communication process. The effects of introducing emotional concerns into the design process of advertising design is also discussed. This study was conducted to develop a new strategy that could provide an optimised communication process, which would essentially include the emotional concerns of the creative and target audience.

Keywords: Emotion · Advertising · Creativity · Communication design

1 Introduction

Interactions of roles in the communication process include the ways in which creative and the target audiences influence each other. Creatives are expected to come up with innovative new ways to attract a greater share of their target audience. In order to equip junior advertising participants with the ability to create innovative ideas, research has proved that emotions, as well as cultural values, are factors that affect the process of advertising communication. The influence of human factors such as emotion concerns, cultural recognition, etc. in the communication design process was has begun to be discussed. However, few studies have investigated how they can enhance creativity and interact with the communication process; hence, more research on their influence on the interactions between the creatives and the audience is necessary. This research study aimed to investigate the relationships between emotion and social recognition to motivate the interactions between the roles in the communication process through a literature review. The investigated concept was then illustrated by an example of an interactive project conducted by a team of local junior advertising participants. The manipulation of introducing emotional concerns into the design process of advertising design would be analysed. This study was conducted with the hope of developing a

A. G. Ho (Ed.): AHFE 2019, AISC 974, pp. 55–64, 2020.
https://doi.org/10.1007/978-3-030-20500-3_6

new strategy which can provide an optimised communication process, essentially including the emotional concerns of creatives and the target audience.

2 Emotion and Mood

Several scholars have proposed that there were relationships between emotions and moods. The terms 'emotion' and 'mood' were first proposed by psychologists. Plato first described 'emotion' in a poem [1]. Plato and his student, Aristotle, adopted the term 'emotion' to describe the thinking processes involved the body, mind, perception, thinking, wants and feelings without clear identification. The interpretation of the Greek scholars was focused on the complex mental activities of the human mind. They regarded 'emotion' as the feeling of being human as well as the way that humans think. Their interpretation of 'emotion' embraced the connection of emotion, physical and mental, perception, thinking and feeling.

The terms 'emotion' and 'mood' represented certain overarching concepts for psychological studies. Even though these two words were mentioned frequently throughout the literature, most scholars apply the concepts of 'emotion' and 'mood' interchangeably. In order words, they agreed that the structure of 'emotion' and 'mood' were connected with personal experience and individual perspective consideration. A study conducted by Beedie, Terry and Lane [2] evaluated how ordinary people understand the differences between the two terms. Based on the literature review, the research team compared various terms those interpreting emotional changes and mental activities socially from the psychological point of view. Next, they invited a total of 160 participants to provide feedback about the selected terms. By applying the content analysis, 16 selected terms, including reason, duration, control, experience, and consequences were identified as the most frequently adopted terms for describing emotions.

Some scholars investigated classic literature in order to understand the meaning of 'emotion' in the concepts of ancient aesthetics. They found that the term was connected with tastes, judgement and essence. Also, the term 'emotion' was sometimes mentioned in the descriptions of criticisms of artwork or music. Tuske [3] investigated the concept of emotion in Indian philosophical literature. He found that emotion was applied to refer to the responses evoked by art and played an important role in Indian literature. This term involved feelings and personal emotions including happiness, joy, disgust, fear and love. This finding proved that emotions were involved in thinking in both Western and Asian cultures. The literature among both Western and Asian cultures reflected that the concept of emotion worked closely with cognition. The concept of emotion significantly influenced the process of cognition when creating rational interpretations and decision-making. Similar concepts were developed in the studies of spiritual philosophy. Emotion has always been related to people's thinking processes. However, compared to Asian concepts, the concept of emotion from a Western point of view and its destructive influence on thinking was emphasised. In a preliminary understanding of the human thinking process, Western scholars divided mental activities into thinking and feeling. They believed that thinking is a rational consideration process for understanding the environment around us. According to this perspective, emotion was regarded as a responsive action towards the external environment.

3 Applying Emotion in Advertising Communication

Changes in the economy, society and technology shaped a new chapter for the world. Following the changing world, communication design followed the changes. It took on the role of helping to give new orders and reconstruct society with a new and modern spirit. It developed from skill-based professions to more generalised knowledge and ability. The new concept of communication design is highly inclusive of various aspects of daily lives. It engages in daily human lives with a multitude of perspectives. Communication in branding was the aspect of communication design studies that were related to audience-centred concerns, as mentioned earlier. Some scholars suggested building up the motivational story during the communication process [4]. They pointed out that senses and emotional changes formed the design experience [5, 6], and also supported creativity that enhanced the sustainability of communication [7]. In other words, the brand works with the experience it creates to satisfy users' needs. The designer should also understand the audience's values sufficiently. These values are delivered through graphics and language used in the communications between brands and their audience.

Understanding customers' needs and attitudes are the criteria for brands to create strong loyalty and repurchasing behaviour [8, 9]. Such brand loyalty is generated from emotional attachment and passion. It is also the main factor that can motivate consumer engagement [10]. Some brands create meaning and emotional attributes with themes as well as symbols in the communication with the audience [11]. Themes were the central concept presented as the theme of the advertising campaigns.

On the other hand, symbols worked as the execution of the applied theme and promise of the brand to the audience. The application of theme and symbols would be found in the advertisements of cosmetic brands. The advertisements of cosmetic brands were not selling products and services but the dream of the audience, as described by Packard [12]. Bernays [13] has a similar concept to Packard. Bernay investigated the influence of the theme on the effectiveness of advertisements in his book. He found that strong themes would provide enhanced motivation for the audience. This motivation was developed based on the audience's subconscious desires for accessing specific goals.

4 Steps for Designing Emotion

Based on the previous investigation in the studies mentioned above, some techniques were explored for stimulating the audience's emotional responses toward brand messages. It was noticed by some scholars that the interactive communication between audiences and brands became more frequent during the previous 20 years under the influence of new media development. Davis [14] pointed out that brands are unable to build particular brand images with deep emotional attachments by only adopting one media form. Instead, adopting multiple media formats would be a more effective approach. Davis explained that the branding communication presented through multimedia provided more touch points to link the audience up with the brand. Through various touch points, brands build a bond with their target audience in areas such as sense, tone and manner. These enhanced the effectiveness of interactive communication for achieving emotional satisfaction and empathy among the audience. Packard proposed effective approaches for inviting the audience to promote as well as create promotional messages for the brands. He pointed out that messages were promoted to audiences through video, graphics, audio, films, etc. Some of the brands cooperated with content created in various media such as books, publications, and television for providing audience experiences. Packard [12] analysed that the customer services team for most brands launch their promotional activities for engaging the audience through social media. The audiences were grouped as followers and processed multi-directional interaction. In this stage, the audience were linked up with brands. Packard [12] summarised some essential trends, analysed the promotional campaigns and proposed several parts of brand communications. He proposed that brand strategies were the framework for developing the communication process based on integrated media marketing in various places. Social media would be the most effective media for execution included providing customer service to the audience. Branded memberships are reward programmes that would group the supporters of brands into 'big audience brands'. Real-time responses towards personalising information would be sufficient to ensure the growth of connections with the audience. After understanding the framework for designing emotion in advertising, there is a need to know how the creatives would provide emotional satisfaction to the audience, how would the creative choose the appropriated touch points as well as approaches for communication? Also, how would they process the design executions?

4.1 Stimulating Emotion

Some scholars studied how emotion could influence the function of communication design. After the audience was given the information, the persuasive procedure could be started. Persuasion is a process of influencing the audience. According to Perloff [15], there were five perspectives to understanding persuasion. First, a communication process needs loud and sufficient messages delivered from one party to another. Second, it creates influence. The creative should understand the audience and what makes them tick, making any persuasive attempt more likely to succeed. Third, persuasion involves many different elements, not only words. The elements of aesthetics, interactions, functions etc. create visual communication in order to be more persuasive to potential audiences. Fourth, persuasion was not solely depended on humans own attitudes and it would lead an audience to do something purposely. Fifth, the attitudes of the audience are strengthened through the process of persuasion. Scholars found that as there are too many opinions, they have to be reminded from time to time to build faith among the audience.

Petty and Cacioppo [16] studied how persuasion works on individuals through shaping attitudes and behaviours by persuasive arguments. They proposed the model of elaboration likelihood, which explains the presence of messages and design maximising the influence of design execution on the attitudes as well as the behaviours of the audience. They proposed that while someone received information, at the same moment certain levels of elaboration were processed. 'Elaboration' means the thinking process of someone in terms of evaluation, memory, and judgement. They suggested that people would elaborate on the information they understood when they received messages in an effective approach. There were two levels of elaboration which were taken by messages during the processing route: the accessible route and 'peripheral route processing'. On the other hand, 'central route processing' refers to the attention of the audience when they received the message. The audience would mostly pay more attention to bold and persuasive messages. The attitudes of the audience were shaped or developed in 'central route processing', which was developed to endure and resist counter-arguments. Compared to 'central route processing'; 'peripheral route processing' was relatively more difficult to recognise. The audience paid relatively little attention to messages when they were being influenced by the other factors – for example, the origin of the source, the visual appeal, the presentation of the information, etc. The attitudes of the audience were shaped. Thus, they would be relatively less influenced by counter-arguments; they needed continually reinforced information about the brand instead. At this stage, 'central route processing' led the audience to have a higher understanding ability than 'peripheral route processing'. As a result, the structures and contents of the brand information were harshly judged by the audience. The audience is also influenced by other factors besides the contents of the message as well. Their attitudes are influenced by the message characteristics, including the strength, credibility and relevancy of the information they obtained. The audience would also possibly be influenced by the effectiveness of other factors. At the same time, they would be more tolerant towards the persuasive messages and less inclined to adopt the counter-arguments.

Many design elements are tailor-made for peripheral route processing. The use of visual hierarchy is the first obvious part. Working as the focal point of the product, a nice, large product photo is the best tool for attracting the attention of the customers. Moreover, multiple angles of product shots were provided. Following this, certain filtering options of the product features with a broad range of categories were provided. This serves as a shortcut for audiences to select the product they are interested in and to encourage them to conduct in-depth research on the price, rating, product features, etc. Ken represents those casual customers who had relatively little motivation to buy anything. He is curious to know if there are any televisions he can buy within his budget. After a rough understanding of products in the 'search bar on the homepage' on the Amazon website, he then sorts the products based on the price levels from low to high. After that, he uses the rating filters (mostly located on the left of the website); he prefers to search the products with four stars or above only. Ken notices the money that can be potentially saved using the low-price guarantee, which is provided together with his purchase. In addition, 'free shipping' is displayed in bold type and is located right next to the price on the website. This presentation of the data, which appeals to the audience's pocket, is a practical approach to peripheral route persuasion. This can remind penny-pinchers that they do not need to pay extra for the convenience and can have the product delivered directly to their front door.

4.2 Evoking Audience Emotion

Both central route processing and peripheral route processing lead to the same influence on the audience, and design elements were not exclusive to one route. The audience often processes information by adopting certain levels of both routes. Hence, they often work together. For example, Susan adopted the information of the product description by using the central route but utilised the filter of the star-rating as a peripheral route. It was also a shortcut to sort televisions highly rated by previous customers who had similar mindsets. Hence, she was persuaded by elements from both routes. Susan maintained her positive attitude towards purchasing on Amazon.com, whereas Ken needed some further convincing to ensure that he did not go down the street and check the 'big box stores'.

However, creatives need to notice the negative side: that persuasion goes together with messaging and design. Sometimes there were distractions that can undermine creatives' persuasive techniques if the potential audience encounters pop-ups, long loading times, or steps that are too complicated for the appropriate messages. The audience would not get to choose the information. These distractions, regardless of whether they were physical, or intangible, would all be discovered during the elaboration process. The two types of consideration supported further investigation into how emotional concerns played their roles during visual communication. Three main elements worked for persuading audience effectively based on the analysis by Petty and Cacioppo [16]. They found that messages, design outcomes and the methods of message delivery are essential criteria for understanding the audience. They also emphasised that creatives had to understand their audience, including which type of brand information would motivate them. The findings of Petty and Cacioppo provided fundamental knowledge for developing a list of questions to understanding the values of the audience with emotional concerns. Creatives can also conduct a brief review of previous research on persuasion in the field, which will contribute to the effectiveness of the design.

4.3 Visualising Emotion

During a similar period, some scholars explored the relationship between design outcomes and emotional changes. Through surveying the emotional changes of the audience, Sauvagnargues [17] and his team investigated how the linkages of the brand products and its audience's emotional reactions would be motivated by the execution during the communication process. The research team summarised their findings and explained that the design execution provided by the brand involved some emotional elements. According to their study, these emotional elements influenced the interpretations of the audience as to their desires. On the other hand, creatively applied visual elements such as colours and shapes, as well as typography, could be used for expressing messages with emotional elements. Thus, there is a need to understand how design execution would evoke the audience's emotional changes.

The shape is one of the most recognisable visual languages. Various types of shapes delivered different messages and interpreted different moods and emotions. For example, a circle mostly delivers relatively positive messages: audiences tend to associate circular graphics with friendship, trust, satisfaction, and team spirit. Triangles are associated with activity and power. Audiences tend to associate triangular graphics with religion and authority. The spaces among the shapes also created messages and emotions.

4.4 Languages of Colours

Colour is one of the most fundamental elements of design. It stimulates audience reactions in ways that may be very different from other audiences. Individual preference, personal feelings or any other factors may cause different reactions. It is worthwhile for creatives to understand how different colours affect different people.

4.5 The Messages of Typefaces

Different typefaces depict different stories. Typography delivers messages for developing strong identities as well as a concrete impression. Typography presents the mood throughout the communication process with different styles. A typeface should be paired up with the appropriate graphics in an advertisement. This was one of the most important criteria for effective visual communication design. It was found that emotion influenced the interpretation of the audience about the message they received. Moreover, they would then judge if they would adopt the message delivered in visual communication. How can design influence the emotion of the audience? How can creatives change the audience's feelings by stimulating visual communication design? How can we creatively organise the emotions/feelings presented by the design elements?

5 Case Study

Emotions were always deployed in branding and advertising design to create persuasive and interactive communication with target audiences. A branding company would continuously empower the emotional bonding with the audience by reviewing various aspects of their daily lives. This would show a kind of understanding of the customers' needs and attitudes, and even their evaluative values, thus leading to the establishment of strong brand loyalty from emotional attachment and passion. FortunePharmHK's (幸福醫藥) 2015 branding television commercial (TVC) 'Lion Rock Spirit' (獅子山精神)[1] [18] was a good example that shows how the brand cleverly deployed emotional elements to represent and satisfy the targeted young audience's needs and attitudes toward Hong Kong core values and identity in the name of the Lion Rock spirit, after a series of youth-led social movements, such as the Umbrella Movement in 2014 [19, 20]. The TVC created meaning and emotional attributes with the theme and symbols of the Lion Rock spirit to articulate an understanding of the younger generation's espousement of Hong Kong core values and identity in contemporary social recognition. The TVC worked with other communication designers to stimulate the young audience's emotional responses toward the brand message that FortunePharmHK is a local brand that embodies the Hong Kong identity, experience, and values in the name of the Lion Rock spirit. Indeed, the TVC was widely shared among the Hong Kong

[1] Lion Rock spirit (獅子山精神) is used to represent Hong Kong's core values and spirit. It is traditionally articulated in terms of the Hong Kong stories of hardship and perseverance in a TV series 'Below the Lion Rock' (獅子山下) produced by RTHK in the 1970s and 1980s.

youth through social media platforms such as Facebook and YouTube. It received more than 140,000 clicks on the official Facebook page in the first two days when it was launched in April 2015 [21]. This kind of branding strategy – adopting multimedia platforms to engage with young audiences through social media – enhanced the effectiveness of brand and interactive communication to achieve emotional satisfaction and empathy among the audience to a great extent.

5.1 The Theme of the Lion Rock Spirit

The theme of the TVC was the Lion Rock spirit that represented the Hong Kong core values and identity by the young generation's discourse of localism, with a strong desire for autonomy and self-determination. The TVC cleverly presented a branding story to support the younger generation's struggle for 'the right to decide their future' [22] without emphasising the conflict between the different generations' articulations of the Lion Rock spirit. Indeed, the older generation's discourse on the Lion Rock spirit was rearticulated to show their emotional understanding of the contemporary social problems that the younger generation was currently encountering. Finally, the local pharmacy brand aligned its local identity and experience with the young audience and encouraged them to think positively and to solve problems together in terms of the Lion Rock spirit, as well as remembering the brand's motto of 'always going one step further' (幸福要走多步).

5.2 The Symbols of Hong Kong Core Values and Identity

Different symbols were employed in the TVC design to effectively communicate with the young audience in order to create an understanding of their needs and attitudes toward cultural values and social recognition in Hong Kong society, thus creating a brand loyalty from their emotional attachment and passion toward Hong Kong values and identity. Graphics and language were used in the advertising communication to symbolise the brand's promise to the young audience.

The opening showed a crowd of youngsters with mirrored images on the water surface, the MTR station of the University of Hong Kong, and a butterfly shot to represent a kind of desperate mood among the younger generation in Hong Kong, especially after the Umbrella Movement. The older characters such as Uncle Ho (何伯), Ip Chun (葉準), Ming Gor (明哥), and Wong Chi-poon (王志本) presented their generation's demonstration of Hong Kong values by their hardships and perseverance.

However, their narratives such as 'I wish this society no longer required Ming Gor' (我想個社會唔再需要明哥), who donated welfare rice boxes to people in need, and 'I've always been down to earth' (我都冇離地) by Wong Chi-poon, who was a professional accountant, with a sad and sorrowful intonation in front of a wall of property advertisements outside an agency, represented the older generation's understanding and empathy toward social problems encountered by the youth in contemporary Hong Kong. The three post-1980 characters including local illustrator, Tse Sai-Pei (謝曬皮), the multimedia celebrity, Tat Gor (達哥), and the innovative entrepreneur and founder of GoGoVan, Steven Lam (林凱源), represented the success stories of the younger generation by their freedom of expression, independence, and creative flexibility. Their stories and narratives encouraged the youngest generation, such as the millennial student Ho Ngai-chi (何藝之), who also represented the Umbrella Movement generation, to fight for the right to decide their future. For example, Tat Gor's speech 'Who says gamers are losers?' (邊個話打機冇出色㗎) was emotionally connected with Ho Ngai-chi's statement 'In fact, we are not naïve.' (其實我哋唔係唔識諗) In addition, a voiceover announced, 'Even knowing there is a great chance of losing, we must still work hard to win' (明知會輸, 我哋都一定要贏) when the screen showed a young man hiking up a fog-shrouded mountain. The visual design symbolised the younger generation's feelings of uncertainty, but the announcement also revealed their uncompromising struggle for the future. At the end of the TVC, the final narrator's statement 'Living in Hong Kong has never been easy; however, every generation has their own "Lion Rock spirit"' (香港條路從來都唔易行, 但每一代人都有佢嘅獅子山精神) was FortunePharmHK's rearticulation of the different generations' interpretations of Hong Kong's core values and identity in harmony for sharing passion and hope to the young audience, which was visually represented by the twilight sea of clouds.

6 Conclusion

Creatives are expected to come up with messages in innovative ways to attract more of their target audience. The interactive response among roles in the communication process includes how creatives and their target audiences influence each other. For equipping junior advertising participants' abilities to generate creative ideas, the influence of emotions as well as cultural values on the process of advertising communication was found. The influence of some human factors included emotion concerns, cultural recognition and so forth were started to be discussed. However, few studies investigated how they can enhance creativity and interactive within the communication process. Hence, their influence on the interactions between the creative and the audience was discovered. This research study aimed to investigate the relationships between emotion and social recognition to motivate the interactions between the roles in the communication process through literature review. It was found that both central route processing and peripheral route processing lead to the same influence on the audience, and design elements were not exclusive to one route. The audience often processed information by adopting certain levels of both routes. Design execution provided by the brand included visual elements such as colours and shapes, as well as typography and language influenced the interpretations of the audience as to their desires and emotional responses. This investigation then was illustrated by some case studies created by local junior advertising participants. The manipulation of introducing emotional concerns into the design process of advertising design would be analysed. As a result, this study explained how advertising would the optimised communication process by including the emotional concerns of the audience.

References

1. Price, A.W.: Emotions in plato and aristotle. In: The Oxford Handbook of Philosophy of Emotion, pp. 121–142 (2010)
2. Beedie, C., Terry, P., Lane, A.: Distinctions between emotion and mood. Cogn. Emot. **19**(6), 847–878 (2005)
3. Tuske, J.: The concept of emotion in classical Indian philosophy (2011)
4. Roberts, K.: Lovemarks: The Future Beyond Brands. Powerhouse Books, New York (2004)
5. Morrison, S., Crane, F.G.: Building the service brand by creating and managing an emotional brand experience. J. Brand Manag. **14**(5), 410–421 (2007)
6. Rahinel, R., Redden, J.P.: Brands as product coordinators: matching brands make joint consumption experiences more enjoyable. J. Consum. Res. **39**(6), 1290–1299 (2013)
7. Lynch, J., De Chernatony, L.: The power of emotion: brand communication in business-to-business markets. J. Brand Manag. **11**(5), 403–419 (2004)
8. Wang, H.X., Chen, J., Hu, Y.C.: The consistency of product design and brand image. In: IEEE 10th International Conference, pp 1142–1144 (2008)
9. Bergkvist, L., Bech-Larsen, T.: Two studies of consequences and actionable antecedents of brand love. J. Brand Manag. **17**(7), 504–518 (2010)
10. Park, C.W., Jaworski, B.J., Maclnnis, D.J.: Strategic brand concept-image management. J. Mark. **50**, 135–145 (1986)
11. Carroll, B.A., Ahuvia, A.C.: Some antecedents and outcomes of brand love. Mark. Lett. **17** (2), 79–89 (2006)
12. Packard, V.: The Hidden Persuaders. Washington Square Press, New York (1957)
13. Davis, J.: Competitive Success; How Branding Adds Value. Wiley, Chichester (2010)
14. Bernays, E.L., Howard, W.C.: The Engineering of Consent. University of Oklahoma, Norman (1955)
15. Perloff, R.M.: The Dynamics of Persuasion: Communication and Attitudes in the 21st Century. Routledge, Abingdon (2010)
16. Petty, R.E., Cacioppo, J.T.: The elaboration likelihood model of persuasion, pp. 129–170 (1986)
17. Sauvagnargues, A.: Deleuze et l'art. PUF (2005)
18. [FortunePharmHK]: FortunePharmHK 2015 TVC – 'Lion Rock Spirit' (幸福醫藥2015電視廣告 - 獅子山精神) [Video File], 30 March 2015. https://www.youtube.com/watch?v=D9IIoRTJ9I0
19. Kwong, Y.: State-society conflict radicalization in Hong Kong: the rise of 'anti-China' sentiment and radical localism. Asian Aff. **47**(3), 428–442 (2016)
20. Lou, J.J., Jaworski, A.: Itineraries of protest signage: semiotics landscape and the mythologizing of the Hong Kong umbrella movement. J. Lang. Polit. **15**(5), 609–642 (2016)
21. Sky Post: City Talk (城市熱話). Sky Post (晴報), p. P08, 2 April 2015
22. Veg, S.: The rise of 'localism' and civic identity in post-handover Hong Kong: questioning the Chinese nation-state. China Q. **230**, 323–347 (2017)

An Investigation into the Power of Digital Media in Hong Kong

Edward C. K. Hung(✉)

Department of Journalism and Communication, Faculty of Arts, Hong Kong Shue Yan University, 10 Wai Tsui Crescent, Braemar Hill Road, North Point, Hong Kong
ckhung@hksyu.edu

Abstract. In view of the surge of digital media and their overwhelming influence over the communication practice in Hong Kong, this preliminary study is to reveal its power through ten digital advertising cases. These cases cover the recent popular technologies used in digital advertising and the preferred ways of communications among the target customers. This paper dissects the cases according to the functional affordances of the discursive interface analysis and concludes that these cases do share a subset of the functional affordances forming the boundary and exclusion in the digital media in Hong Kong subtly and affecting the designing of the next generations of these digital media.

Keywords: Digital advertising · Digital media · Functional affordances

1 Introduction

In the recent years, there is a surge in the usage of digital media in Hong Kong [1]. Beauty Exchange, Facebook, HKET.com, Google, Instagram, and YouTube are some of the popular ones [2]. Meanwhile, the advertising dollars spent on Hong Kong digital media is on the rise [1]. Mobile video, native content for smartphones, in-app advertising, and programmatic advertising are the focus. Hongkongers have been living in the digital age. Are they using digital media to get what they want (i.e., communications, advertising, learning, etc.)? Or, they are being controlled by digital media. Understanding the relations between digital media and their users affect the future design of digital media. In this paper, we try to reveal such relations in Hong Kong through digital advertising on digital media.

Hung and colleagues [3] claim that it is possible to study digital advertising on digital media to grasp the nature and "behaviors" of digital media. In this paper, we refer to the functional affordances of the discursive interface analysis [4] to dissect these digital advertisements in order to reveal the power of the relevant digital media. In other words, our hypothesis is that the digital media in Hong Kong is influencing Hongkongers in certain ways. We will carry out a preliminary study of 10 cases of digital advertising campaigns in Hong Kong to test this hypothesis. We would like to highlight that since digital advertising on digital media is heavily dependent on the Internet and the discursive interface analysis is specially designed to analyze the interfaces on the Internet through sensory, cognitive, and functional affordances, using

A. G. Ho (Ed.): AHFE 2019, AISC 974, pp. 65–70, 2020.
https://doi.org/10.1007/978-3-030-20500-3_7

functional analysis to dissect digital advertising is reasonable and appropriate. Herein, sensory affordances are the design choices of websites, such as color and interactivity, enabling their users to sense them. Cognitive affordances are the linguistic and nominative features of websites, such as names and labels, informing their users of their capabilities. Functional affordances are the norms of websites defining what they can do. In this preliminary study, we redefine functional affordances as *the marketing and advertising rules, models or frameworks adopted by a digital advertising platform including social media and website that allow a digital advertising campaign to happen in a particular way.*

2 Methodology

The hypothesis in this paper is that the digital media in Hong Kong are influencing Hongkongers through its power that can be revealed in the functional affordances of the digital advertising campaigns in Hong Kong. This paper adopts qualitative analysis on ten of these digital advertising campaigns in the period between 2011 and 2018 to perform a preliminary study to test the hypothesis.

3 Ten Cases of Digital Advertising Campaigns in Hong Kong

In this section, we study ten cases of digital advertising campaigns in Hong Kong. Based on the related articles and our analyses, we define the functional affordances used by them as shown in Table 1.

Table 1. The functional affordance of the ten cases of digital advertising campaigns in Hong Kong

	Companies	Case descriptions	Functional affordances
1	Fujifilm [5]	Fujifilm Instax used "Photo of Photo" posts on social media to tell stories in Hong Kong and to target hipster audience	Consumer engagement, community experience, medium specificity, product trial, social networks, storytelling, target segment, user experience
2	Ocean Park [6]	Ocean Park Hong Kong engaged its customers through celebrity, Key Opinion Leader (KOL), viral video, 360° panoramic video, social media, online advertisements, search engine marketing (SEM), and tailor-made mobile app to offer a better tour experience	360° panorama, consumer engagement, cross-platform mobile app, social networks, storytelling, Search Engine Marketing (SEM), target audience, user experience
3	China Mobile [7]	To acquire younger customers, China Mobile adopted a social media campaign strategy to share a	Consumer engagement, social networks, storytelling, target segment

(*continued*)

Table 1. (*continued*)

	Companies	Case descriptions	Functional affordances
		micro movie starring a popular YouTuber in addition to the mini games on Facebook	
4	The Hong Kong Jockey Club [8]	To provide a positive horseracing image and acquire younger customers, the Hong Kong Jocky Club organized "Digital Nights at Happy Wednesday" to let racegoers experience the jockey's eye-view of a race and feel the thrill of sitting on a fast-moving horse through the Samsung VR Gear. Racegoers could upload their photos taken with the "Master Jockey" frame after reaching certain score level to the carnival's Facebook page to compete for round trip air tickets to Seoul	360° panorama, hashtagging, interactive game installation, social networks, target segment, user experience
5	Wyeth Nutrition [9]	Wyeth Nutrition offered an educational app with Google Street View to soft sell its products. The app enabled children to visit famous tourist attractions such as Big Ben and the Taj Mahal through a 360° virtual tour while learning the pronunciations of the attractions in English, Cantonese, and Mandarin. The children could also earn "miles" after using the app to redeem coupons and other prizes	Consumer engagement, cross-platform mobile app, target audience, user experience
6	Coca-Cola [10]	To have a younger and energetic brand image, Coca-Cola HK urged its fans to shake their mobile phones equipped with Coca Cola's shake-to-play mobile game when a Coca Cola TVC was played. The players then entered a lucky draw to win prizes from Coca-Cola	Consumer engagement, cross-platform mobile app, medium specificity, target segment, user experience
7	China Construction Bank [11]	China Construction Bank (Asia) organized Hong Kong Wine & Dine Festival for brand building. The festival used a giant digital social wall, namely Happy Moment Social Wall to display its Facebook and Instagram photos of happy moments uploaded by its participates to	Consumer engagement, hashtagging, interactive kiosk, medium specificity, social networks

(*continued*)

Table 1. (*continued*)

	Companies	Case descriptions	Functional affordances
		encourage social sharing and photo uploading	
8	Henderson [12]	Henderson adopted an integrated digital solution, namely Samba Foodlympics to promote Sunshine City Plaza and Shatin Plaza. Sitting on a large banana boat and wearing a pair of VR glasses programmed for two motion-detection-enabled paddles, a visitor could interact with MR. MEN and LITTLE MISS in a VR rowing game. To play the game, the visitor had to spend a specific amount in one of the plazas and register on the campaign site	Consumer engagement, interactive game installation, medium specificity, social networks, target segment, user experience
9	IKEA [13]	To celebrate its 40th anniversary, IKEA offered a social media campaign to let participants (i) vote and share their favourite stories behind IKEA's top 10 classics on Facebook to win prizes; (ii) join a "snap & win" Instagram photo contest to share their daily lives with IKEA's products and designated hashtags to win mega rewards	Consumer engagement, hashtagging, social networks, target audience
10	Optical 88 [14]	To have electronic word of mouth, Optical 88, an eyewear brand, organized a social CRM campaign featuring 88 Bear as the brand mascot. The participants who logged into Facebook, watched videos about eye care tips & tests, shared the videos, invited friends to join, and uploaded a photo with the 88 Bear at Optical 88 could win prizes	Brand mascot, consumer engagement, Customer Relationship Management (CRM), hashtagging, social networks

4 The Power of Digital Media in Hong Kong

Based on the ten cases in Table 1, 64.7% (11/17) of the listed functional affordances are shared. Two of them are used by eight cases. They are consumer engagement and social networks. That is to say, most of the cases aimed at engaging the target customers through social networks such as YouTube, Facebook, and Instagram. On the other hand, there are three functional affordances used by four or five cases. They are medium specificity, target segment, and user experience. It is possible to claim that these companies were very eager to convert new technologies into new media to communicate with their target customers and to offer them a unique experience. In addition, there are six functional affordances used by two or three cases. They are 360° panorama, cross-platform mobile app, hashtagging, interactive game installation, storytelling, and target audience. Those functional affordances that were not shared among the cases are brand mascot, community experience, CRM, interactive kiosk, product trial, and SEM.

In this preliminary study, the power of digital media in Hong Kong is revealed in those functional affordances. They form the boundary of these Hong Kong digital media. They support and define the features of these digital media for communication purposes. Under these circumstances, their users are also limited or even controlled by them, affecting the ways they communicate with the others. The hypothesis in this paper is supported. Meanwhile, these functional affordances form the exclusion of these digital media. Any digital media in Hong Kong without these functional affordances might find themselves difficult to enter the market since their target users have been very used to those digital media with the shared functional affordances.

5 Conclusion

The functional affordances of the ten cases in Table 1 hint that when designing the next generations of the digital media in Hong Kong, we must put these affordances into considerations since they define and limit the related features. For example, should we take out the third-party apps function of Facebook [15] when designing its next generation? Does this violate the functional affordances called consumer engagement and product trial? Understanding the power of digital media does help the designing of their next generations.

To conclude, we study ten cases of digital advertising campaigns in Hong Kong to reveal the power of the relevant digital media in the form of functional affordances. We also reveal that these affordances affect the designing of the next generations of these digital media. In the future, we will expand our study with more cases from different places to verify our initial findings in this paper.

References

1. He, L.: Hong Kong's Digital Spending to Surge to US$5.8b by 2022 as Consumers Turn to Mobile Media (2018). https://www.scmp.com/business/companies/article/2149582/hong-kongs-digital-spending-surge-us58b-2022-consumers-turn
2. Marketing-interactive: Hong Kong's Top Digital Media – Media Report 2018. https://www.marketing-interactive.com/hong-kongs-top-digital-media-media-report-2018/
3. Hung, E.C.K., Chan, A., Chan, R.: A study of rules, structures, features, and biases in the new media for digital advertising for the development of new media ontology. In: International Conference on Applied Human Factors and Ergonomics, pp. 88–92. Springer, Cham (2018)
4. Stanfill, M.: The interface as discourse: the production of norms through web design. New Media Soc. **17**(7), 1059–1074 (2015)

5. New Digital Noise: NA. Fujifilm: Achieving 30X More Organic Reach on Social Media. https://newdigitalnoise.com/portfolio/fujifilm
6. Ocean Park Hong Kong: Three New Features on the Ocean Park Hong Kong App (2018). https://www.oceanpark.com.hk/en/park-information/news/three-new-features-ocean-park-hong-kong-app%20
7. New Digital Noise: NA. Providing 360° Digital Communications Solutions to Transform China Mobile Hong Kong. https://newdigitalnoise.com/portfolio/china-mobile
8. The Hong Kong Jockey Club: Happy Wednesday (2016). https://www.facebook.com/happyweds/
9. Spikes Asia: See the World at Home (2013). https://www2.spikes.asia/winners/2013/mobile/entry.cfm?entryid=1930&award=101&order=2&direction=2
10. TNW: The Awesome Coca-Cola Campaign Shows the Vast Potential of Mobile Marketing (2011). https://thenextweb.com/shareables/2013/03/12/this-awesome-coca-cola-campaign-shows-the-vast-potential-of-mobile-marketing/
11. Marketing-interactive: Case Study: Hong Kong Wine & Dine Festival 2015 (2015). http://www.marketing-interactive.com/features/case-study-hong-kong-wine-dine-festival-2015/
12. Pixo Punch: Henderson O2O Solution & VR Experience: MR. MEN LITTLE MISS SAMBA FOODLYMPICS (2016). https://www.youtube.com/watch?v=1IsjwUcVtbo
13. Marketing-interactive: Ikea Invites Fans to have a Sleepover (2013). https://www.marketing-interactive.com/ikea-invites-fans-sleepover/
14. Chan, J.: Optical 88 Keeps an Eye on You with New Campaign (2015). https://www.marketing-interactive.com/optical-88-keeps-an-eye-on-you-with-new-campaign/
15. Hung, E.C.K.: An exploration of the creative work ontology for the ontology aggregation of social media. Commun. Media Asia Pac. (CMAP) **2**(1), 37–50 (2018)

Understanding How Advertising Gamification Influences Consumers: The Effect of Image Experience and Interactivity

Yunbo Chen and Huijie Yao(✉)

Jinan University, Guangzhou, China
35308063@qq.com, 961243856@qq.com

Abstract. In recent years, advertising gamification has considerably evolved as an approach to increase consumer engagement and motivation. Given the continuous development of game interactivity and the multi-dimensional technology, the potential of advertising gamification still needs to be fully understood. The aim of this study is to investigate how advertising gamification influence consumers' advertising attitude, brand image and brand attitude by looking into the effect of different image experience (2D/3D) and interactivity (interactive/non-interactive). This paper conducts an experiment (N = 40) among young people aged 20–26 years old to compare the effects of advertisements. The main effects are significant, which provide strong evidence of how an interactive gamified experience and higher media image experience can increase advertising attitude, brand image and brand attitude, and provide new insights about the effectiveness of several game mechanics. Ultimately, we can also explore the importance of interactive design in the process of advertising gamification and the possibility of its application in the fields of AR.

Keywords: Advertising gamification · Interaction design · Image experience · Brand attitude

1 Introduction

Research on communication tend to focus on one emerging medium. However, technology is constantly evolving, making the research results likely to be outdated. Therefore, this research attempts to explore more than a single medium, examining the common morphological characteristics of the development of media (PC, mobile, and VR). For example, stronger interactivity, upgraded visual experience, and a sense of immersion brought by the two are consider common features that can be observed during the development of different media, particularly in the process of gamification.

The concept of gamification was first put forward in 2002 by Nick Pelling, a British computer programmer. Sebastian Deterding defined the notion as "the application of randomly combining elements of game design in non-game situations" [1]. Since 2010, the concept began to spread on a global scale. Since gamification makes mundane things interesting and attractive, the concept was first used in pedagogy. Due to the increasing popularity of gamification, the concept of intrinsic motivation was then

A. G. Ho (Ed.): AHFE 2019, AISC 974, pp. 71–85, 2020.
https://doi.org/10.1007/978-3-030-20500-3_8

introduced by Malone [2]; its effect on human cognition was later verified. Malone's theory has since been widely adopted in the field of gamification. In recent years, gamification has become a trend in business and marketing. The phenomena is thought to be caused by multiple factors, including cheaper technology, personal data tracking, the existing success of the practice, and the popularity of games [1]. In addition, gamification was applied in advertising by big companies such as Microsoft and Nike [3]. Although, scholars have paid attention to the role of gamification in improving user experience, advertising gamification has not been thoroughly researched. In China, "gamification marketing" is often used to describe advertising gamification. Increasingly, gamification is used in online retail as a marketing strategy. Understanding the attributes of gamification marketing activities (GMAs) is crucial to effective gamification. Chen (2018) studied a total of 242 online bookstore customers, and found that GMAs has a significant and positive impact on both hedonic and utilitarian value [4].

Many people associate advertising gamification with product placement, in-game advertising, and advergames. In-game advertising is the advertising of products or brands in video games, providing players with product and brand information [5]. Advergame are interactive games specially designed for the purpose of advertising. The relationship between advergames and the brands they advertise is exclusive. Capella (2013) reviewed research on in-game advertising, advertising, and advertising in games in social media. The researchers took players' personal and social factors into consideration. A framework was thus established to distinguish the stimulating properties of games, players' psychological responses to games, and their actions toward games and brands, paving a way for future research on digital game advertising [6]. Both in-game advertising and advergames are forms of advertising that incorporate the design of games based on digital media. However, games existed long before the emergence of digital media. Games are, fundamentally, a form of medium. Gamification should be considered an advertising strategy and independent from other forms of media. However, gamification can be mixed with different media, generating new forms of advertising gamification. When print media was the dominant form, advertisements attracted attentions not only by copywriting and images, but also by games such as crossword. The use of games in print media was to increase audience interaction and can, thus, be described as gamification. However, print media limit game options and how they are presented. Therefore, little attention was paid to gamification in print media. In the current era of digital media, the booming gaming industry has led to the popularity of advertising gamification. Hence, the concept has been closely tied to digital media. According to the above analysis, game placement and advergames are merely the manifestations of advertising gamification in digital media. Furthermore, advertising gamification should have universal designs and applications that can be applied in different media forms. Audience's reception of advertisement should be redefined; audience participation should be emphasized; and audience interaction should be enhanced. Finally, the effectiveness of advertisements should be further optimized.

The design of interactive games serves as a guidance in advertising gamification. Media technology is the foundation for the development and popularization of advertising gamification. While games are intrinsically attractive, players demand for higher level of participation and immersion. This is the reason why AR and VR technologies are first applied in games. As a driving force, media technology also increases participants' immersion level. Cohen (2016) proposed the concept of Avatar Identification [7], which stems from character identification. The difference is that Avatar Identification projects the desire to become a certain character onto an avatar. When avatar identification is at work, an avatar replaces an individual's real self. On this basis, through enhanced image experience, players generate avatars, which are influential and empathic with the real self.

In 2017, China's PC and mobile gaming markets witnessed an upward trend, with a market size of 5.06 billion yuan and 148.92 billion yuan [8]. In addition to large enterprises, independent studios also showed outstanding performance. Game production chain was refined, laying a foundation for the popularization of advertising gamification and the development of its technology in China. Since businesses are increasingly aware of the significance of advertising gamification in improving advertising effectiveness, important questions arise: how to design the right interactive games in the right media; and whether incorporating more senses in advertising experience will cause adverse effects. The lack of theories, however, makes the issues remain unsolved.

This study focused on advertising gamification. The research samples included university students, post-graduate students, and white-collar employees. Situational experiments and in-depth interviews were used. From the angle of game interaction and visual experience, the research designed and utilized four Ikea advertisements as materials to find out audience's attitudes towards advertising, brand image, and brand attitudes when game interaction is or is not used. Furthermore, the research studied the effects brought by the use of 2D or 3D images (or both). The importance and possibility of immersive experience and game interaction in advertising gamification were also explored.

2 Literature Review

2.1 Design of Game Interaction

James (2014) believed that replacing static public signs with digital screens creates opportunities for the application of interactive display systems. The systems can be used in collaborative workspaces, social gaming platforms, and advertising. With the help of marketing communication theories and the existing consumer behavior models, the research analyzed the effectiveness of interactive display advertising based on its three stages, namely attraction, interaction and integration [9]. Smart mobile devices should be integrated into interactive displays to increase their effectiveness as advertising tools. The process of designing game interaction is the foundation of the interaction between audience and advertisements; however, it also limits the ways of interaction. To put it simple, the design of game interaction is guided by the concept of "games" and driven by interaction design. The two elements perform their separate functions, making interactive game advertising not just a tool to attract audience and deliver information, but also have interaction and integration functions. Regarding research on audience's memory of the implanted information in video games, as interactivity leads to more detailed and deeper processing, Dardis and Auty use processing fluency theory to explain why interactivity makes information more memorable. However, according to persuasion knowledge model, consumers will gradually become aware of the marketing intentions and the strategies used by the market to achieve the expected results, and gradually become immune. Then, the ability of consumers to identify persuasion intentions will gradually develop into the ability to treat advertisements with suspicion and rationality. However, add game interaction design in advertisements so that the lack of clear advertising clues or "bumpers" in games may reduce the vigilance of consumers and transfer their favorable feelings towards games to the advertisements they watch. Therefore, the increase of game interaction design in advertising will have a positive impact on advertising, brand recall, and brand attitudes.

2.2 Image Experience

Through experiments on 60 university students, Li et al. (2002) compared the experience brought by 3D and 2D online advertisements and discussed the characteristics of 3D advertising and its impacts on consumers. Li also attempted to verify that the presence of 3D online advertisements is more powerful than that of 2D online adverts [10]. Most of the materials for experiments on 2D and 3D advertising were from the Internet, belonging to the field of digital advertising. In-game advertising and advergames were also introduced in the Internet era. However, compared with traditional mass media such as magazines and TV, the interactivity of the Internet is greatly enhanced due to input devices such as mouse and touch screen. Although the revenue of online advertising ranks the highest, outdoor adverts are booming thanks to AR and VR technologies. If the effects of image dimensions are only compared between online advertisements, it will be difficult to draw a conclusion with high applicability. Therefore, the subjects of this paper are not limited to online advertisements. 2D and 3D adverts are selected to compare their influences on audience.

2.3 Influence of Advertising Gamification: Advertising Attitudes, Brand Impression and Brand Attitudes

Based the hierarchical effect model, which includes unconsciousness, consciousness, knowledge, preference, belief and purchase, Hogg et al. (2004) extended the scale of the process after user interactions. The study revealed that there exists a progression of consumers' attitudes between their first contact with advertisements and the act of purchasing. Ralf et al. (2013) built a research framework for advertising gamification, examining audiences' psychological responses after being influenced by brands and games. Ralf also explored audiences' final actions towards brands and games. The characteristics, responses and behavioral options of the three phases were listed. In this research, dependent variables for the experiment were extracted based on Ralf's framework, including advertising attitudes, brand impression and brand attitudes.

3 Research Hypotheses

Advertising attitudes are defined by consumers' cognition and emotions towards advertising, with advertising value and advertising intrusion as its influencing factors. (Logan 2012) Advertising value is determined by consumers' evaluation of advertising content based on its informativeness, entertainment level, and offensiveness; advertising intrusion refers to the consumers' psychological response to the form of advertising that is disruptive to their continuous cognitive process [11]. It is hypothesized that the entertainment level of the advertisement will prompt a positive response from the consumers, improving consumers' impression on the advertisement by increasing the advertising value. Edwards (2002) had concluded before Logan that the advertising intrusion of pop-up style online advertising could be mitigated by increasing its advertising value [12]. The informativeness and entertaining level of the advertisement also have an adverse effect on advertising intrusion and positive effect on advertising attitude. Moreover, Coyle et al. (2001) had discovered that interactivity influences the attitude towards online advertising [13]. The above conclusions are reached via information system research. Therefore, there exist fewer empirical studies on the influence of interactivity on advertising attitude. Within these studies, Campbell (2008) conducted two experiments on the influence of interactivity and personal correlation on advertising attitude which showed that advertising interactivity has a significant influence on people's attitudes towards online advertising [14]. This paper bases its research on the aforementioned studies and concludes that interaction games will have a positive impact on consumers' advertising attitude.

H1a: Interactive games have a positive impact on consumers' advertising attitude.

If the advertising presentation, which is the media image experience, can maintain a smooth and continuous advertising cognition for consumers and reduce advertising intrusion, it can improve consumers' advertising attitude. Kerrebroeck et al. (2017) verified that advertising in the form of 3D virtual reality (VR) aroused a more positive attitude than 2D videos due to its vividness and presence [15]. Johanna et al. (2018) compared the influence of 2D, 3D, and VR versions of the same video game on the player's experience and the brand placements [16]. Since the game interactivity varies in different medium, players of the same games enjoy a different level of entertainment. The entertaining level of advertising has an important impact on consumers' advertising attitude. Therefore, it is believed that the higher the immersion of the medium, the more positive the advertising attitude.

H1b: The immersion experience of media image has a positive impact on the advertising attitude.

Impression refers to the feeling stirred by objects exposed to a person [17]. Brand impression refers to the part or the whole brand information left in consumers' mind by their perception and experience of brand information. Impression changes as audience is exposed to new and objective information. Therefore, the formation of impressions is a never-ending, continuous process. There are two main views regarding the information integration theory of brand impression formation. One of them is Anderson's accumulative model and average model (1968) [17], which is of the view that the audience evaluates the meaning of each piece of information of the target object, and then integrates the evaluations according to laws of algebra. On the other hand, Asch (1946) proposed that the audience integrates various characteristics of the target object into a single, logical impression. However, he further proposed that the degree of impression-causing stimuli varies between the central traits and peripheral traits. [18] Therefore, when studying the influence of gamified advertisement on brand impression, it is necessary to realize that the stimulation effect of a single advertisement builds on the audience's existing brand impression, and the resulting effect may be integrated and absorbed. When measuring the variable, emphasis should be placed on the degree of the elevated or declined impression after exposure to the advertisement, to explore whether a gamified advertisement is effective in building a positive brand impression.

Beuckels et al. (2016) measured respondents' perception of luxury goods through interactive and non-interactive experiments in virtual reality scenes. The result shows that due to a heightened sense of presence, the image interactivity leads to higher exclusivity, quality, enjoyment and self-expanding perception. [19] Therefore, this paper proposes that interactive games that are conducive to pleasant experiences is effective in building a positive brand impression.

H2a: Interactive games are effective in building a positive brand impression.

Hoffman and Novak (1996) conducted a series of research and model development on flow theory. Attention is the precursor of flow. Whether the audience can immerse into the environment is the premise for flow interaction. The immersive feeling brought by 3D media image experience is more intense than 2D media image experience. Therefore, this paper proposes that media image experience in a higher dimension has a positive impact on audience's brand impression.

H2b: The media image experience in a higher dimension has a positive impact on audience's brand impression.

Long-term advertising campaign aims to build a positive attitude towards brands. Attitude is complex as its basic components include cognition, emotion and behavioral intention. Any changes in these components will lead to changes in attitude. Kara Chan (1996) tested consumers' responses to rational and emotional advertisings. The results showed that consumers preferred emotional advertisements that are attractive, interesting and creative. Advertisements that emphasize products' objective properties such as ingredients, performance, packaging and price are regarded as monotonous, boring and easy to forget. Therefore, in order to attract audience and establish a good brand image, the integration of emotional elements in advertising is crucial. The effects caused by emotional advertisements, such as attractiveness, amusement and creativity, are similar to that of gamified design and yet they have subtle differences. Games are naturally interesting and attractive. The interactive experience and media immersive experience provided by gamifying advertisement may enhance audience's brand attitude. Hence, this paper proposes that interactive games in advertising have a positive impact on brand attitude.

H3a: Interactive games in advertising has a positive impact on brand attitude.

Creative advertising that is conducive to emotion is more attractive to the audience than rational advertising. In the gamification of advertising, the direct purpose of designing interactive games is to enhance the audience's experience of watching advertisement through entertainment from the games. The higher media dimension provides the audience with a fluid and smooth experience. Choi (2014) investigated the influence of 3D advertising on online shopping, concluding that the vividness of psychological intention can act as a medium that could adjust the 2D and 3D effects, directly affecting consumers' attitudes and intentions. It is observed that the effect of 3D advertising is better than 2D advertising [20]. Hence, this paper proposes that the higher media dimension has a positive impact on improving brand attitude.

H3b: Higher media dimension has a positive impact on improving brand attitude.

4 Research Design and Data Collection

4.1 Research Subjects

Forty people aged 20–26 (20 males and 20 females) were selected as experimental subjects. Their occupations include university and graduate students (50% of subjects) students and fresh graduates who are currently working in white-collar professions (50% of subjects) The subjects were asked whether they had any knowledge about the brand IKEA, and rated their brand preference (between 0 to 5 points) prior to the test. It was shown that all subjects had had a good understanding of IKEA and had seen IKEA's advertisements with a 4–4.5 preference rating without deviation. It indicated that all subjects belong to IKEA's target consumer group and have a similar preference for IKEA.

4.2 Design of Experiment

The experiment selects and designs four advertising materials with two sets of 2 advertisements with manipulated variables. The experiment settles on 2 (interactive design without game vs. interactive design) * 2 (2D media experience vs. 3D media experience) to conduct a contrastive experiment. The 40 subjects had been randomly assigned to four groups, with 10 people in each group. They completed quantitative questionnaires after inspecting the advertising materials in their group. The purpose of the experiment is to detect audience's responses to advertising attitude, brand impression, and brand attitude after interacting with the gamified advertisement and media experiences with different dimensions. In order to ensure the consistency of the process and enhance the reliability of the experiment, all 40 subjects were guided by a researcher when completing the experience and evaluation (Table 1).

Table 1. Experimental grouping

Group	Interactive design (with or without games)	Dimension of the media image	N	%
1	Interactive design (without game)	2D	10	25%
2	Interactive design (with game)	2D	10	25%
3	Interactive design (without game)	3D	10	25%
4	Interactive design (with game)	3D	10	25%

The experimental procedure is as follows:

1. After the list of subjects has been confirmed, the subjects were randomly sorted into four groups prior to the experiment. The researchers required the subjects to inspect the respective advertising materials and then fill in the questionnaire regarding the advertisements they have just experienced. The respective advertising materials for each group are shown in Fig. 1.
2. Control of the interactive game and media dimensions.
   (1) The two groups inspecting the 2D advertising saw two posters with household products and IKEA logo. The poster without interactive game only features product graphics and copy-writing. Posters with interactive game feature two picture with a few differences. The copy-writing prompts the subjects to look for the differences between the two pictures and win a prize by complying to the request of interaction.
   (2) For the two groups experiencing the 3D images needed to wear VR glasses to watch panoramic videos. The group experiencing advertisement without interactive game feature were shown scenes from IKEA physical store. The group experiencing advertisement with interactive game feature required the usage of an AR application. After a demonstration from researchers, the subjects could operate the application by themselves. They could move and display IKEA household products in a virtual reality scene, and change the colors and style of the products.
   Due to the significant differences between the four experiment materials, a set time for the experience is not imposed. Instead, the subjects determined when they have finished the experience and informed the researchers to conduct the next step, the questionnaire on the effect of the advertising experience.
3. After the experiment is completed, researchers recorded the questionnaire results into spss22, and conducted follow-up telephone calls with subjects whose data revealed significant deviation from the normal range to discover the underlying reasons for the deviation.

4.3 Dependent Variable Measurement

Advertising Attitudes. To measure advertising attitudes, a 5-point semantic differential scale is adopted. The five points ranging from −2, −1, 0, 1, 2 representing the negative to positive experiences which measure by five categories (good/bad, pleasant/unpleasant, attractive/unattractive, enjoyable/unenjoyable, and interesting/boring) related to advertising attitude in previous researches [21].

Brand Impression. According to past researches, the stimulation effect of advertising builds on the audience's original brand impression. The final effect might be integrated and absorbed by the original band impression. Therefore, the focus of the measurement was put on the degree and direction of the after-effect the materials had on the audience's brand impression, in order to measure whether gamified advertising is able to positively influence brand impression.

Brand Attitudes. Brand attitude is defined as an internal assessment by an individual on a brand. [22] With this definition, the complex attitude can be measured by a semantic differential scale. There are four categories on the scale measuring brand attitude (Anand et al. 1990), including good/bad, like/dislike, pleasant/unpleasant, and enjoyable/unenjoyable [23].

5 Data Analysis and Hypotheses Testing

5.1 Reliability Analysis

The scale used in this study is a modification of that of the previous research. Therefore, reliability analysis is required to validate the scale. Cronbach alpha is used for the analysis. The analysis results are shown in Table 3.

The Semantic Difference Scale of advertising attitude and brand attitudes was translated according to the measurements proposed by Spears [21]. Hence, the scale can comprehensively measure the subjects' attitudes after viewing the materials. With two items' Cronbach alpha value higher than 0.8, the scale is deemed reliable. Brand impression was measured by the Richter scale to find out whether viewing advertising materials leads to positive impression. Regarding brand impression, the Cronbach alpha value is higher than 0.6, indicating acceptable reliability. Overall, the Cronbach alpha value for the questionnaire is above 0.8, suggesting that the questionnaire is reliable and consistent.

5.2 Hypothesis Testing

To test the hypotheses, for each of the dependent variables, independent sample variance analysis was conducted to compare the means between different conditions (see Tables 2 and 3).

Table 2. The experiment materials of each group

	Interactive design (without game)	Interactive design (with game)
2D	 Group 1	 Group 2
3D	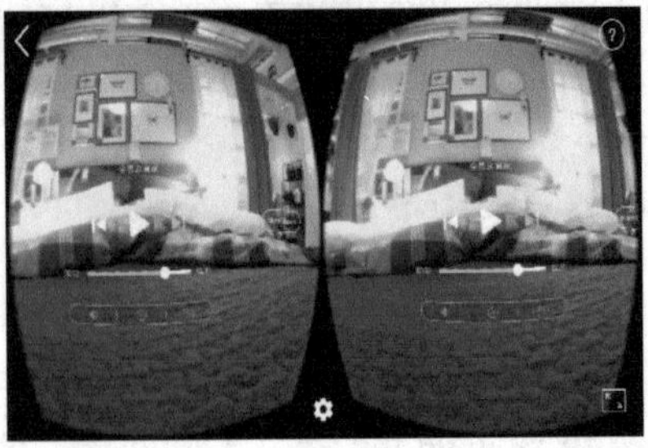 Group 3	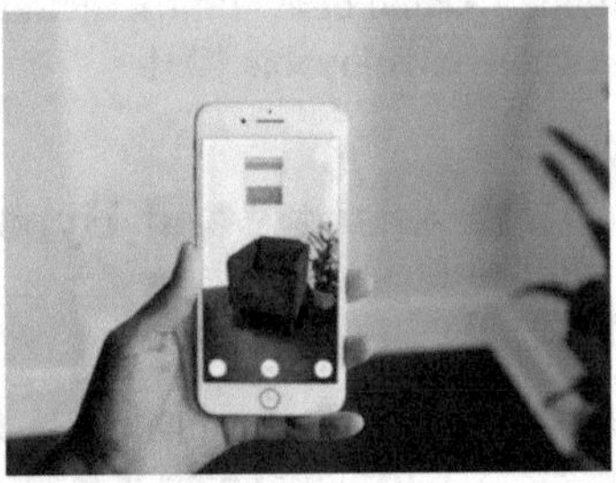 Group 4

Table 3. Reliability analysis results

Scale items	Cronbach Alpha	Numbers
Advertising attitude	0.879	5
Brand impression	0.618	3
Brand attitude	0.879	5
Overall	0.911	13

The scale of the semantics differences between advertising attitudes scored a continuous value of −2 to 2. This study used variance analysis to examine the results.

In order to test H1a and H1b, this study used game interaction design and image dimensions as independent variables to conduct a 2 * 2 variance analysis of advertising attitude. The results show that the design of game interaction has a significant positive impact on the subjects' advertising attitude: M (with games) = 1.15 > M (without games) = 0.40, $F(1, 40) = 7.70$, $P < 0.01$. In addition, the results also suggest that image dimension has a positive impact on advertising attitude: M (3D) = 1.25 > M (2D) = 0.30, F (1, 40) = 12.35, $P < 0.01$. Hence, H1a and H1b are supported.

From the results, it can be found that game interaction design and image dimension have no significant two-dimensional interaction effects on advertising attitude: F (1, 40) = 0.034, p = 0.854. That is, the positive effect of game interaction design on advertising attitude in 2D media is no weaker than that in 3D media. According to the interviews with samples od high deviation, the main reason may be that 3D media remains immature. Some people experienced slight dizziness when watching 3D videos; some experienced difficulties when using the AR app. The comfort level of the subjects decreased when watching 3D videos, affecting their attitudes towards advertisements. Therefore, the increase of image dimension does not strengthen the positive impact brought by game interaction design on advertising attitudes.

Since brand impression was measured by the Richter scale coded from −3 to 3, with continuous scores, this study used variance analysis to examine the results. In order to test H2a and H2b, this study used game interaction design and image dimension as independent variables to conduct a 2 * 2 variance analysis of brand impression. The results show that game interaction design has a significant positive impact on the brand impression of the subjects: M (with game) = 1.45 > M (without game) = 0.40, F (1, 40) = 5.313, $P < 0.05$. In addition, the results suggest that image dimension has a positive impact on brand impression: M (3D) = 1.40 > M (2D) = 0.45, F (1, 40) = 4.349, $P < 0.05$. Therefore, H1a and H1b are proven correct.

From the data results, we find that game interaction design and media dimensions have no significant two-dimensional interaction effects on brand impression: F (1, 40) = 0.012, p = 0.913. That is, the positive effects brought game interaction design on brand impression in 2D media is no weaker than that in 3D media. According to the interviews with samples of high deviation, the bright colors and lively styles of Group1 and 2's materials affected the subjects' objective evaluation of brand impression as they prefer cool-toned colors. Similar problems also existed in the group that viewed 3D advertisements. The main reason is that the subjects gave scores according to the contents of the materials; while the independent variables of the experiment are the technology and design of the advertising. Although, the study attempted to make the contents of the four advertisements consistent, it is difficult for the contents to satisfy each subjects' preference. Therefore, the positive influence of game interaction design on brand impression does not increase with the increase of media dimensions.

Since brand attitudes were measured by the Richter scale coded from −3 to 3 and the scores were continuous numbers, this study used variance analysis to examine the results. In order to test H3a and H3b, this study used game interaction design and media dimensions as independent variables to conduct a 2 * 2 variance analysis of brand impression. The results show that game interaction design has a significant positive impact on the brand attitudes of the subjects: M (with game) = 1.00 > M (without game) = 0.05, F(1, 40) = 4.661, $P < 0.05$. In addition, the results suggest that the dimensions of media has a positive impact on brand attitudes: M (3D) = 1.00 > M (2D) = 0.05, F (1, 40) = 4.661, $P < 0.05$. Hence, it is safe to conclude that H1a and H1b are correct.

From the results, we find that game interaction design and media dimensions have no significant two-dimensional interaction effects on brand attitudes: F (1, 40) = 2.182, p = 0.148. That is, the positive effects of game interaction design on brand attitudes in 2D media is no weaker than that in 3D media. According to the interviews with the samples with high deviation, they believe that the two 3D ads have a stronger presence. It is also safe to conclude that the innovative technologies employed by companies can result in positive attitudes towards brands. The main reason may be that brand attitudes are formed indirectly through watching advertisements, and are affected by advertising attitudes, which are formed earlier on. Hence, the positive influences of game interaction design on brand attitudes are not strengthened by the increase of media dimensions (Tables 4 and 5).

Table 4. Dependent measure across different conditions (game interaction design)

Measure	Interaction design (with game)	Interaction design (without game)	Degrees of freedom	p<
Advertising attitudes	1.15	0.40	7.70	0.01
Brand impression	1.45	0.40	40	0.05
Brand attitudes	1.00	0.05	4.66	0.05

Table 5. Dependent measure across different conditions (media image experience)

Measure	3D	2D	Degrees of freedom	p<
Advertising attitudes	1.25	0.30	12.35	0.01
Brand impression	1.40	0.45	4.35	0.05
Brand attitudes	1.00	0.05	4.66	0.05

6 Conclusion and Discussion

This study used experiments to find out the effects of advertising gamification, game interaction design (with/without) and media dimensions (2D/3D) on audience's advertising attitudes, brand impression and brand attitudes. It was found that both game interaction design and media dimensions have significant positive effects on the improvement of all these three aspects.

Game interaction design enhances advertising attitudes, brand impression and brand attitudes. The study was partly inspired by propositions from the literature and developed these propositions further with concrete evidence. For example, it was argued in literature that game interaction design can increase the entertainment of advertisements. During the experiments, subjects spent more time watching advertisements with game interaction design and paid more attention comparing to those without. In addition, game interaction design successfully lured the audience away from the mere persuasive nature of advertising and caught them off-guard when feeding advertising information along with the game content, which was conducive to increasing the attitude toward the advertisements. Game interaction design also due to higher exclusivity and helped the audience to better perceive brand information, thus improving brand impression. At the same time, the emotional attributes in game interaction design such as entertainment and exciting content can improve the brand attitude of the audience when compared to traditional rational advertising.

Media dimensions is also effective in promoting advertising attitudes, brand impression and brand attitudes. Higher sense of immediacy is a major advantage of 3D advertising. The sense of presence and immersion brought by multiple dimensions of media can reduce advertising intrusion and create a more immersive flow, improving the viewing experience of advertising, hence advertising attitudes. In terms of enhancing brand impression, multi-dimensional media creates the same effect as game interaction design, enabling the audience to perceive brand information more attentively. Finally, from in-depth interviews with the subjects, a new factor associated with brand attitudes improvement was found – the novelty. As media develop, the improvement of image dimensions not only represents more advanced technology, but also brings along refreshing innovative audio-visual experiences. During the interviews, more than half of the subjects mentioned that the application of new technology often reflects the comprehensive strength and innovation capability of a company, which helps enhance the attitude toward the company, i.e. the brand attitude. And subjects who had less or even no previous exposure to 3D videos and AR technology rated higher on brand attitudes toward the two brands with 3D advertisements. High novelty means that the public has not yet got bored of this medium and developed immunity, that is to say, media with more dimensions produce stronger effects on the audience. Essentially, game interaction design and media dimensions are two paths to the same destination – to improve advertising effect: the former disguises the persuasive nature of advertising with integrated design, whereas the latter enhances the persuasive effect by offering better technology-empowered experience. According to the results of the experiments, these two approaches both have positive impacts on improving advertising attitudes, brand impression and brand attitudes.

7 Limitations and Recommendations

In order to ensure the external reliability of the experiment, the pre-experiment of brand attitudes was conducted by sacrificing the internal reliability, where the initial impressions of attitudes toward IKEA's advertisements and the brand were measured. The improvement of brand impressions and attitude improvement was measured as well, so that the subjects could understand the purpose of the experiment. Secondly, the experimental design simplified the environmental factor and did not include the social factors into the covariate analysis. Without mimicking the social and environmental context, the conclusions might not be comprehensive enough. Thirdly, some of the experimental material were newly produced, while some others were carefully selected from IKEA advertisements. Although the study avoided using advertisements that are too distinctive and have strong message to maintain the style of all advertisements consistent and similar to the style of the products advertised, the results would have been more credible if all the experimental materials were newly produced to eliminate the influence of the interference factors.

After the completion of this experimental study, we found that there is still much to explore about advertising gamification. For example, social factors can be integrated to produce more representative results. The study selected subjects from the same age groups with an even distribution of men and women. However, the age groups were limited by the target age groups of IKEA. If brands or products with a wider range of audience are selected in the future, it is possible to study the impacts of game interaction design and media dimensions on different age groups. Secondly, in the course of the experiments, because of the different material forms and the lack of time control of each experience, it was noticed that participants who had longer experience time tended to have stronger willingness to participate. The effects of the two maneuverable variables on the audience's advertising experience time could be examined in the future. Moreover, during follow-up studies for subjects with skewed results, it was found that their previous exposure to AR also contributed to their attitudes toward advertising. Those who had previous exposure would form technology immunity and have higher expectations for advertising. Therefore, they showed less improvement of attitudes or brand impression upon watching the ads comparing to those who had never been exposed.

References

1. Deterding, S.: Gamification: designing for motivation. Interacts. J. **19**(4), 14–17 (2012)
2. Malone, T.W.: What makes things fun to learn? A study of intrinsically motivating computer games. J. Pipeline **6** (1981)
3. Gamification Introduction. https://www.zhihu.com/question/20381247
4. Chia-Lin, H., Mu-Chen, C.: How gamification marketing activities motivate desirable consumer behaviors: focusing on the role of brand love. J. Comput. Hum. Behav **88**, 121–133 (2018)
5. Yang-jing, Li-xianguo, Wang-chao.: The Influence of the Frequency of Game Placement on the Effect of Brand Placement – Based on the Moderating Effect of Game Involvement and Interactivity. J. China Soft Sci. **10** (2016)
6. Terlutter, R., Capella, M.L.: The gamification of advertising: analysis and research directions of in-game advertising, advergames, and advertising in social network games. J. Advertising **42**(2–3), 95–112 (2013)
7. Cohen, J.: Audience identification with media characters. Psychol. Entertain. J. **1**, 183–197 (2016)
8. China mobile game industry report. http://report.iresearch.cn/report_pdf.aspx?id=3266
9. Crowcroft, J., et al.: Convergence of interactive displays with smart mobile devices for effective advertising: a survey. J. ACM Trans. Multimedia Comput. Commun. Appl. **10**(2), 1–16 (2014)
10. Li, H., Daugherty, T., Biocca, F.: Impact of 3-D advertising on product knowledge, brand attitude, and purchase intention: the mediating role of presence. J. Advertising **31**(3), 43–57 (2002)
11. Logan, K.: And now a word from our sponsor: do consumers perceive advertising on traditional television and online streaming video differently? J. Mark. Commun. **19**(4), 258–276 (2013)

12. Edwards, S.M., Li, H., Lee, J.H.: Forced exposure and psychological reactance: antecedents and consequences of the perceived intrusiveness of pop-up ads. J. Advertising **31**(3), 83–95 (2002)
13. Coyle, J.R., Thorson, E.: The effects of progressive levels of interactivity and vividness in web marketing sites. J. Advertising **30**(3), 65–77 (2001)
14. Campbell, D.E., Wright, R.T.: Understanding the role of relevance and interactivity on customer attitudes. J. Electron. Commer. Res. **9**, 62–76 (2008)
15. Van Kerrebroeck, H., Brengman, M., Willems, K.: When brands come to life: experimental research on the vividness effect of virtual reality in transformational marketing communications. J. Virtual Reality **21**, 177–191 (2017)
16. Johanna, R., Ralf, T., Stefano, T.: The same video game in 2D, 3D or virtual reality - how does technology impact game evaluation and brand placements? J. PLOS ONE **13**(7), e0200724 (2018)
17. Jin, S.: Social Psychology. Higher Education Press, Beijing (2010)
18. Asch, S.E.: Forming impressions of personality. J. Abnorm. Soc. Psychol. **4**, 303–314 (1946)
19. Beuckels, E., Hudders, L.: An experimental study to investigate the impact of image interactivity on the perception of luxury in an online shopping context. J. Retail. Consum. Serv. **33**, 135–142 (2016)
20. Choi, Y.K., Taylor, C.R.: How do 3-dimensional images promote products on the internet? J. Bus. Res. **67**(10), 2164–2170 (2014)
21. Spears, N., Singh, S.N.: Measuring attitude toward the brand and purchase intentions. J. Curr. Issues Res. Advertising (CTC Press) **26**(2), 53–66 (2004)
22. Mitchell, A.A., Olson, J.C.: Are product beliefs the only mediator of advertising effect on brand attitude? J. Mark. Res. **18**, 318–332 (1981)
23. Anand, P., Sternthal, B.: Ease of message processing as a moderator of repetition effects in advertising. J. Mark. Res. **17**(August), 345–353 (1990)

Locality and Local Identity Discourses in Post-handover Hong Kong Brand Advertisements

Sunny Sui-kwong Lam(✉) and Terry Lai-sim Ng

The Open University of Hong Kong, Hong Kong, China
{ssklam, tlsng}@ouhk.edu.hk

Abstract. Since the handover in 1997, identity politics have arisen in Hong Kong social and political movements. Locality and local identity discourses have become unprecedentedly salient to rearticulate Hong Kong culture and identity. People's feelings, thoughts and behaviors in relation to community solidarity and identity construction in the name of "bentu" or locality are significantly advocated among Hongkongers in terms of the sense of belonging and emotional bonding. Similarly, brand advertising deploys emotional appeals to build brand trust and loyalty among target consumers. Brand advertisements have increasingly deployed much more complicated multimodal semiotics and metaphors to create identity myths to emotionally affect and persuade consumers by the sharing of collective representations of local cultures and identities. This study will investigate how Hong Kong advertisers deploy multimodal semiotics in their brand advertisements to establish brand equity in line with the recent cultural emotions and sentiments of Hongkongers.

Keywords: Belongingness · Branding strategy · Emotional affections · Identity discourses · Locality · Multimodal semiotics

1 Introduction

Since the handover in 1997, Hong Kong social and political contexts have been dramatically changed, thus leading to the advent of identity politics. Locality and local identity discourses have become unprecedentedly salient to rearticulate Hong Kong culture and identity. The further challenge of the differentiation of Hong Kong identity from the national identity by Beijing's assimilation project arouses the rise of localism and new types of local identity discourses. People's feelings, thoughts and behaviors in relation to community solidarity and civic identity in the name of "bentu" (本土) or locality are significantly advocated among Hong Kong people, especially the young generation, to define their collective beliefs and actions for autonomy and self-determination [1, 2]. Likewise, brand advertising and promotional strategies also deploy emotional appeals of locality to build brand trust and loyalty among local consumers [3]. Brand advertisements have increasingly deployed multimodal semiotics and metaphors to create identity myths to emotionally affect and persuade consumers by the sharing of collective representations of local cultures and multi-faceted

A. G. Ho (Ed.): AHFE 2019, AISC 974, pp. 86–94, 2020.
https://doi.org/10.1007/978-3-030-20500-3_9

identities. The emotional locality branding strategies need to consider not only the local language but also the local thoughts as the loci of multimodal semiotics and metaphors onto the brand positioning and resonance with the consumers' identities and lived experiences.

This study will investigate how Hong Kong advertisers deploy multimodal semiotics and metaphors in their brand advertisements to construct the brand identity in line with the recent cultural emotions and sentiments of Hongkongers. In addition, the metaphors need to emotionally map with the locality and local identity discourses for the young generation's flexible identifications. Nonetheless, the brand advertisements still need to persuade the other audiences of local and pan-Chinese cultural values and identities in Hong Kong to identify with the brand and its products in order to build positive brand associations. Three case studies of local and global corporates will be investigated to discern how their brand advertisements deploy locality and local identity discourses to emotionally affect their consumers by flexible identifications.

2 Locality and Local Identity Discourses

"Identities are never unified and, in late modern times, increasingly fragmented and fractured" and "never singular but multiply constructed across different, often intersecting and antagonistic, discourses, practices and positions" in modern societies [4]. Identity is a symbolic construction among people in an imagined community and the identity construction is fluid and intangible [5, 6]. The act of identification is also a continuous process to represent a community identity such like "Hongkonger" (香港人) based upon the interrelationships among subjects, subject-positions, and discursive practices. The (imagined) local community identity is symbolically constructed through collective representations and practices in traditions shared by a group of people to generate an overall sense of belonging through a coherent and stable process of identification [4]. However, identity is a construct that is "replete with paradoxes" and the paradox "combines notions of sameness and continuity with notions of difference and distinctiveness" in a multifaceted context [6, 7]. Therefore, in addition to tradition, identity may be constructed through differences, especially under the circumstances of identity politics. This differentiation from others favors the rise of localism and new types of local identity discourses by flexible identifications. This phenomenon of identity politics has arisen in Hong Kong since the handover in 1997 and the corresponding local identity discourses have become increasingly critical and antagonistic after a series of youth leading social movements[1].

Locality, or bentu, redefines the Hong Kong identity through discourse about civic and local identity in a new mode of identification under a strong context of mainland-Hong Kong antagonism [1, 2]. The Hong Kong local identity is a duality based on "collective beliefs about shared attributes, values, and experiences" among communities in Hong Kong [8]. The young generation tends to rearticulate the identity of

[1] The series of social movements include the Anti–Express Rail Link movement in 2010, the Anti–National Education campaign in 2012, the Umbrella Movement in 2014, and the Mong Kok riot in 2016.

Hongkongers by their flexible identifications through collective and connective actions [9]. The local identity (or identities) of Hong Kong people has never been unified; however, the rise of localism and new local identity discourses reflect the antagonism and increasingly strong fragmentation in society. This is revealed through the two different discourses of Lion Rock spirit (獅子山精神)[2], which reflect and represent the core values and shared attributes and experiences among different generations of Hongkongers, especially after the Umbrella Movement in 2014.

The rise of the Hongkonger identity highlights a "need for differentiation of the self and assimilation with others in their group identification" under the identity politics caused by the anti-mainlandization discourse in Hong Kong [10]. This anti-mainlandization discourse is relatively emotional responses to Beijing's top-down assimilation discourse that tends to enforce the ideology of a Chinese national identity. Therefore, the identity of Hongkongers is reinforced by localism to rearticulate and rebrand "Hong Kong's identity and uniqueness" through discourses on Hong Kong's locality and core values, and the flexible identifications of being localists [1, 10]. The rise of localism and new local identity discourses with a strong claim for autonomy and self-determination by the young generation of Hongkongers provide new challenges and opportunities to identity-based emotional advertising and branding strategies by Hong Kong advertisers.

3 Belongingness and Emotional Advertising

Emotions are strategically deployed in contemporary advertising. The application of emotional appeals in advertising and brand communication elicits contagion effects on consumers' feelings of products and creates a favorable attitude towards a brand [3, 11]. Consumer emotion management and emotional selling proposition (ESP) are emphasized in advertising and branding strategies as "emotions play an important role in decision making" and consumers' "feelings play an important role in the formation of attitudes and judgements about advertisements" and the corresponding brands [12]. Especially positive emotions generate more positive reactions to the advertisements [13] and higher levels of brand recall [14]. However, sometimes, negative emotions can arouse much stronger emotional responses from consumers by their feelings in terms of personal experiences, collective actions and identification in a community which represent a (necessary) sense of belonging. Certainly, appropriation and articulation of these negative emotions to result in positive affective feelings and motivations is the role of advertising. Emotional advertising makes use of intangible, deeply affective

[2] Lion Rock spirit is an ill-defined ideology and discourse to represent Hong Kong core values and spirit. The old generation tends to articulate it with the Hong Kong stories of hardship and perseverance in a TV series "Below the Lion Rock" (獅子山下) produced by RTHK in the 1970s and 1980s. The young generation does not agree with the traditional discourse of Lion Rock spirit but rearticulates it with local democracy and civic identity after the emplacement of a banner that read "I want true universal suffrage" (我要真普選) on Lion Rock during the Umbrella Movement on October 23, 2014. Further information about Lion Rock spirit can refer to the Wikipedia: https://en.wikipedia.org/wiki/Lion_Rock_Spirit.

expressions and impressions "to affect customers' reactions to advertisements, to enhance their attention and to affect brand attitude" [3]. Emotional appealing advertisements are more likely to be remembered compared with rational informative advertisements (Page, Thorson and Heide, 1990) because emotions "are strongly correlated to attention, decision making and memory" [14, 15].

Brand identity and consumer identity are co-created by symbolic interactions among the brand, the individual consumers, and the brand community. Brand identity is a set of unique brand associations and symbols the advertisers use to identify the brand to consumers by means of "the symbolic, visual, and physical representation" [16]. However, brand identity is "more than just visuals", and every brand has its personality just like an individual person [17]. Consumers' individual identity is constructed through individual personality and their shared beliefs, evaluations, values, norms, and social relationships that are constantly changing and shaped by their experiences and symbolic interactions [6, 16]. Identity is a concept of dynamicity that makes every individual consumer feel "a sense of oneself as a coherent and stable entity" and provides consumers "a social process of categorizing" themselves as similar to certain social group members of a community and different from others [6]. When consumers continuously meet and interact with a brand in brand advertisements, emotional affections on them may create a favorable attitude towards the brand leading to positive brand reputation that often sublimates into brand loyalty [17, 18]. Brand community is formed over time as identity co-creation by the reciprocal relationships between brand and consumers [16]. Consumers' feelings of belongingness explain their motivation and behavior to form and maintain a minimum quantity of intimate and interpersonal relationships that emotional branding strategies address [19, 20].

4 Emotional Branding Strategies from Local to Global

Under the highly competitive and matured global markets, cognitive arguments based on rationality are not effective, as it has become very difficult to distinguish one product from another [3]. Therefore, an ESP is advocated by advertisers to engage consumers with a conversation by emotional bonding in order to establish the unique, intangible associations with particular products or brands [21, 22]. In brand advertising, emotional appeals and ESP strategically connect a brand of a human quality with its consumers' innermost feelings, thus satisfying their need to belong by intimate and emotional relationships [23].

Branding strategy is the purposeful presentation and representation of a brand in a strategic plan to generate a positive brand image and identity in the minds of the consumers [17]. An effective brand strategy makes the brand identifiable and is directly connected to the consumers' needs and emotions [24]. Two key factors formulate branding strategy. They are positioning and resonance; both are correlated to consumers' feelings and emotional responses. "Positioning is the perception consumers have" of the projection of a brand and its products, while resonance "makes consumers feel something" by a brand's external, usually emotional, message that "connects with internal values and feelings" of the consumers [25]. Emotional branding strategy creates a brand community by positioning and resonance to make an emotional connection with consumers' identity and feelings of belongingness to provide them with an opportunity to feel like a part of the brand [24]. However, the perception and understanding of a brand's positioning as well as the connection with consumers' feelings to achieve resonance are under control by the consumers' mind. Fortunately, emotional branding strategies can create brand myths of "an emotional context, which provides the platform from which consumers find a sense of identity and belonging" [25].

Not only do local corporates but also global corporations make use of locality and local identity discourses in their brand advertisements. Their television commercials (TVCs) using multimodal semiotics and metaphors of locality and local identities "construct the glocal identity" to reveal some "patterns of glocalization" by means of discursive cultural forms and identities [26, 27]. In Lam's comparative study of the TVC promotion of the Halloween Festival by Hong Kong Disneyland and Ocean Park, both local and global cultures and identities were deployed in their branding and promotional strategies [28]. The use of local cultures and identities in brand advertising is regarded as a strategy of "cultural branding" through which local or global "brands successfully become cultural icons by performing identity myths that individuals draw from to construct their own identities" by emotional bonding [29, 30]. Certainly, "careful planning, designing, and selection of words and visuals" in brand advertisements are important to transform consumers' cultures and identities for positive brand associations as well as "to grant local consumers an emotional membership" in the brand community [31].

5 Identity Myth in Hong Kong Brand Advertisements

Three emotional branding advertisements in post-handover Hong Kong are studied to demonstrate how the local and global corporates deploy locality and local identity discourses to create emotional affections on their consumers. Their TVCs sophisticatedly articulate and rearticulate local cultures and identities by multimodal semiotics and metaphors. In 2015, the local pharmacy Fortune's (幸福醫藥) TVC "Lion Rock Spirit"[3] mythologized the different generations' articulations and rearticulations of Lion Rock spirit to represent the Hong Kong local identity and core values from the past to the present and to connect emotionally the local brand and consumers in a harmony. The China-based, translocal domestic appliance retailer Suning (蘇寧) produced a branding TVC series "He's a Suning man" of three episodes in 2017[4]. Each episode presented a conversation between a young Hong Kong boxer and an old local actor and contextualized the story with the Hong Kong landscapes and cultural symbols to make the brand Suning become a cultural icon by performing identity myths in local context.

[3] Please refer to the link in YouTube for Fortune's 2015 TVC "Lion Rock Spirit": https://www.youtube.com/watch?v=D9IIoRTJ9l0.

[4] Please refer to the links in YouTube for Suning's branding TVC series in 2017. Episode 1: https://www.youtube.com/watch?v=DYms4HDBXFI&list=PL73vGTjgCD58KaSCrtERgG1Ar-xgUSwMH&index=2&t=0s, episode 2: https://www.youtube.com/watch?v=2TpoOiNScs4&list=PL73vGTjgCD58KaSCrtERgG1Ar-xgUSwMH&index=5, episode 3: https://www.youtube.com/watch?v=fVmrFcWUJzo&list=PL73vGTjgCD58KaSCrtERgG1Ar-xgUSwMH&index=3.

Lastly, Nike Hong Kong Limited simply but sophisticatedly deployed some Hong Kong athletes, youngsters and places to glocalize the brand by emotionally mapping with the locality and local identity discourses in 2017[5]. Within all these TVCs, the advertisers deployed emotional branding strategies to generate positive brand associations to affect the young consumers' attitudes toward the brands and to co-create positive brand identities with the consumers' perceptions through the symbolic, visual, and physical representations of locality and local identities of Hongkongers. Table 1 lists the metaphors of locality and representative Hongkongers in the TVCs, which helped to create each brand's identity myths that the youngsters can draw from to construct their own identities to confirm a sense of belonging to the identity of Hongkonger, as well as the brand community.

Table 1. Metaphors of locality and people of local identity in three post-handover Hong Kong brand advertisements.

Brand (year)	Metaphors of locality	People of local identity
Fortune (2015)	Lion rock, martial art, cityscapes and landscapes, MTR station, classroom, rice box	- Old generation: Uncle Ho, Ip Chun, Ming Gor, Wong Chi-poon - Young generation: Tse Sai-pei, Tat Gor, Steven Lam, Ho Ngai-chi
Suning (2017)	Episode 1: Hong Kong ferry, Victoria Harbour, cityscape, Hong Kong film Episode 2: Rooftop, cityscape Episode 3: Antique car, landscape, cityscape	- Rex Tso (all three Episodes) - Tik Lung (Episode 1) - Bryan Leung (Episode 2) - Tsang Kong (Episode 3)
Nike (2017)	Landscapes, cityscapes, playgrounds	- Rex Tso - Cecilia Yeung - Youth athletes

Fortune's TVC "Lion Rock Spirit" used Lion Rock, as well as other metaphors of locality, to symbolize Hong Kong core values and spirit shared by different generations of Hongkongers. Instead of showing the generations' antagonism by their different local identity discourses, the TVC presented a brand of locality to share the brand's local identity and experiences with the young generation in order to encourage them to think positively and to solve the Hong Kong problems together by following the brand's motto of "always going one step further". Uncle Ho (何伯), Ip Chun (葉準), Ming Gor (明哥), and Wong Chi-poon (王志本), who were born in 1937, 1924, 1952 and 1971 respectively, represented the older generation of the local identity and experiences shared with Fortune to withstand the hardships by perseverance and tolerance. However, these old men were sympathetic toward the young generation and agreed with their accusations of the contemporary society and social problems to a

[5] Please refer to the link in YouTube for Nike's 2017 branding TVC "Down But Not Out" in Hong Kong: https://www.youtube.com/watch?v=46sml-dRaDo&list=PL73vGTjgCD58KaSCrtERgG1Ar-xgUSwMH&index=7.

certain extent. Tse Sai-pei (謝曬皮), Tat Gor (達哥), and Steven Lam (林凱源), who are the post-1980s Hongkongers, represented the younger generation who treasured to express their locality and local identities of post-material values through autonomy and self-determination. Their successful stories in line with the local pharmacy brand's story provided a hope to the youngest generation of Hongkongers represented by millennial student Ho Ngai-chi (何藝之). Fortune's TVC rearticulated the old and young generations' discourses of Lion Rock spirit in a harmony by sophisticatedly presenting the reciprocal relationships of the eight characters by emotional bonding of their changing local identity discourses for Hongkongers.

Suning's TVC series "He's a Suning man" deployed the legend by the local super-flyweight boxer Rex Tso (曹星如) of a nickname "The Wonder Kid" (神奇小子), who has won many regional titles and represented the Hong Kong spirit of "never giving up", to inscribe the human quality and emotional essences of responsibility (episode 1), perseverance (episode 2), and ambition (episode 3) onto the brand's identity. He's a Suning man but also a representative of the young Hongkongers. In the three episodes, he had a conversation with each of the three Hong Kong actors: Tik Lung, Bryan Leung, and Tsang Kong, who represented the old generation and the collective memories of Hong Kong film and television. Their conversations showed the different articulations of the meanings of responsibility, perseverance, and ambition in a man's quality with the changing locality. They symbolized the different living attitudes and experiences by the young and old generations but demonstrated the reciprocal relationships among them by their innermost feelings and resonance. For instance, Tik Lung repeated his film dialogue "I'm not a Big Brother anymore" in "A Better Tomorrow" (英雄本色), and hugged and asked for Tso to call him "a neighborhood" (街坊), thus symbolizing the intimate and reciprocal relationship between the two generations of Hongkongers.

Nike's TVC in its brand campaign "Down But Not Out" also deployed the legend of Rex Tso, but Cecilia Yeung (楊文蔚) – the local high jump champion, and other local athletes of both ladies and gentlemen were also presented to tell the local story in line with Nike's motto of "Just Do It". The fighting spirit and the struggling emotions shown by the boxer and other sports players in the TVC in combination with the two taglines: "You can't guarantee a win" (贏唔贏無人知); "But you can guarantee a fight" (搏到盡我話事) strongly aligned with the recent youth generation's identity politics of autonomy and self-determination in Hong Kong. Interestingly, this branding campaign was created by Wieden + Kennedy, Shanghai, China. The TVC did not show any mainland-Hong Kong antagonism, but presented the identity myths of locality through the landscapes, the cityscapes, the playgrounds, and the athletic identities. Nike's brand identity was localized by the emotional bonding with the young, local athletes.

6 Conclusion

Emotional locality branding strategies have played an increasingly important role in the construction of brand identity myths to create the intimate and reciprocal relationships with the consumers leading to the brand community of strong consumer loyalty.

Therefore, emotional advertising using the symbolic, visual, and physical representations of locality and local identities can more effectively affect consumers' reactions to brand advertisements, enhance their attention and affect their attitudes toward the brand. Three cases of brand advertisements in post-handover Hong Kong has been studied to see how the local and global brands localize and glocalize their brand identity in line with locality and local identity discourses. Their TVCs reveal their careful planning, designing, and selection of words and visuals in brand advertisements to transform the consumers' cultures and identities for positive brand associations. Further studies of the consumers' perceptions and reactions to such kind of post-handover Hong Kong brand advertisements of emotional appeals in terms of locality and local identity discourses should be conducted to discern the long-term contagion effects on the constitution of brand associations and consumer loyalty.

References

1. Kwong, Y.: State-society conflict radicalization in Hong Kong: the rise of 'anti-China' sentiment and radical localism. Asian Aff. **47**(3), 428–442 (2016)
2. Veg, S.: The rise of "Localism" and civic identity in post-handover Hong Kong: questioning the Chinese nation-state. China Q. **230**, 323–347 (2017)
3. Panda, T.K., Panda, T.K., Mishra, K.: Does emotional appeals work in advertising? the rationality behind using emotional appeal to create favorable brand attitude. IUP J. Brand Manage. **10**(2), 7–23 (2013)
4. Hall, S.: Introduction: who needs 'identity'? In: Hall, S., du Gay, P. (eds.) Questions of Cultural Identity, pp. 1–17. Sage Publications, London (1996)
5. Anderson, B.: Imagined Communities: Reflections on the Origin and Spread of Nationalism. Verso, London (1983)
6. Scott, S.: Negotiating Identity: Symbolic Interactionist Approaches to Social Identity. Polity Press, Cambridge (2015)
7. Lawler, S.: Identity: Sociological Perspectives, 2nd edn. Polity Press, Cambridge (2014)
8. Brewer, M.B.: Multiple identities and identity transition: implications for Hong Kong. Int. J. Intercult. Relat. **23**(2), 187–197 (1999)
9. Bennett, W.L., Segerberg, A.: The logic of connective action: digital media and the personalization of contentious politics. Inf. Commun. Soc. **15**(5), 739–768 (2012)
10. Yew, C.P., Kwong, K.: Hong Kong identity on the rise. Asian Surv. **54**(6), 1088–1112 (2014)
11. Hasford, J., Hardesty, D.M., Kidwell, B.: More than a feeling: emotional contagion effects in persuasive communication. J. Mark. Res. **52**(6), 836–847 (2015)
12. Kemp, E., Bui, M., Chapa, S.: The role of advertising in consumer emotion management. Int. J. Advert. **31**(2), 339–353 (2012)
13. Goldberg, M.E., Gorn, G.J.: Happy and sad TV programs, how they affect reactions to commercials. J. Consum. Res. **14**, 387–403 (1987)
14. Hamelin, N., Moujahid, O.E., Thaichon, P.: Emotion and advertising effectiveness: a novel facial expression analysis approach. J. Retail. Consum. Serv. **36**, 103–111 (2017)
15. LeBlanc, V.R., McConnell, M.M., Monteiro, S.D.: Predictable chaos: a review of the effects of emotions on attention, memory and decision making. Adv. Health Sci. Educ. **20**(1), 265–282 (2015)
16. Black, I., Veloutsou, C.: Working consumers: co-creation of brand identity, consumer identity and brand community identity. J. Bus. Res. **70**, 416–429 (2017)

17. Drewniany, B.L., Jewler, A.J.: Creative Strategy in Advertising, 11th edn. Cengage Learning, Wadsworth (2014)
18. Cova, B., White, T.: Counter-brand and alter-brand communities: the impact of Web 2.0 on tribal marketing approaches. J. Mark. Manage. **26**(3/4), 256–270 (2010)
19. Baumeister, R.F., Leary, M.R.: The need to belong: desire for interpersonal attachments as a fundamental human motivation. Psychol. Bull. **117**(3), 497–529 (1995)
20. McLeod, S.A.: Maslow's Hierarchy of Needs (2017). https://www.simplypsychology.org/maslow.html
21. Law, J. (ed.): A Dictionary of Business and Management, 6th edn. Oxford University Press, Oxford (2016)
22. Taylor, A.K.: Strategic Thinking for Advertising Creatives. Laurence King Publishing Ltd., London (2013)
23. Moriarty, S., Mitchell, N., Wells, W.: Advertising & IMC: Principles and Practice, 10th edn. Pearson Education Ltd., London (2015)
24. Stec, C.: Brand Strategy 101: 7 Essentials for Strong Company Branding. https://blog.hubspot.com/blog/tabid/6307/bid/31739/7-components-that-comprise-a-comprehensive-brand-strategy.aspx. 11 October 2017
25. Altstiel, T., Grow, J.: Advertising Creative: Strategy, Copy, Design, 4th edn. Sage Publications, Thousand Oaks (2017)
26. Dash, A.K., Patnaik, P., Suar, D.: A multimodal discourse analysis of glocalization and cultural identity in three Indian TV commercials. Discourse & Commun. **10**(3), 209–234 (2016)
27. Robertson, R.: Glocalization: time-space and homogeneity-heterogeneity. In: Featherstone, M., Lash, S., Robertson, R. (eds.) Global Modernities, pp. 25–44. Sage, London (1995)
28. Lam, S.S.K.: 'Global corporate cultural capital' as a drag on glocalization: Disneyland's promotion of the Halloween Festival. Media Cult. Soc. **32**(4), 631–648 (2010)
29. Holt, D.B.: How brands become icons: the principles of cultural branding. Harvard Business School Press, Brighton (2004)
30. Kluch, Y.: Welcome to the peer? national identity, German belonging, and the Abercrombie & fitch brand as social imaginary in reunified Germany. J. Pop. Cult. **50**(6), 1168–1183 (2017)
31. Dash, A.K., Patnaik, P., Suar, D.: A multimodal discourse analysis of glocalization and cultural identity in three Indian TV commercials. Discourse Commun. **10**(3), 209–234 (2016)

Mainstreaming Culture Jamming? Revisiting the Socio-Cultural Impact of Viral Campaign

Chi-kit Chan[1(✉)] and Anna Wai-yee Yuen[2]

[1] The Hang Seng University of Hong Kong, Shatin, Hong Kong, China
chikitchan@hsu.edu.hk
[2] Hong Kong Baptist University (Shek Mun Campus), Shatin, Hong Kong, China
annay@hkbu.edu.hk

Abstract. This paper revisits the socio-cultural impact of culture jamming, a practice which stems from social movement to derail mainstream values and corporate branding. Culture jamming aims to convey messages of social resistance by adding satirical and alternative cultural meaning to the icons representing corporate brands and social mainstream values. However, in view of the rise of social media networks and viral campaigns, culture jamming could paradoxically be incorporated by authorities and corporations for their own promotional purposes. By conducting textual and content analysis of a viral campaign in Hong Kong—a fictitious icon promoting life-saving by the Fire Services Department of Hong Kong government, this paper discusses the conditions which facilitate such mainstreaming incorporation of culture jamming. The study hopes to foster scholarly dialogue on the versatility of culture jamming across the fields of social movement and business branding.

Keywords: Culture jamming · Mainstreaming · Viral campaign · Social media networks · Hong Kong

1 Introduction

This paper explicates how the rhetoric strategy of culture jamming, which stems from the ethos of social movement to disrupt hegemonic discourses, is adapted by social authorities and corporations for their own promotional purposes. Culture jamming has been widely deliberated in literature of social movement and cultural studies for the contemplation of its cultural resistance and alternative social imaginary [1–3]. It is commonly manifested in parodies, satires, identity formation and body politics against political figures, corporate brand, and the mainstream 'common sense' prevailing in societies. In brief, the discursive narration of such mainstream values is reconstructed and derailed from its original cultural meaning [3, 4]. Typical examples include modifying corporate logos into playful icons, jamming pop-songs with political grievances, and name-calling of prominent leaders.

The rhetoric strategy of culture jamming, however, has been revisited by some scholars for its plausible effect of brand promotion and formation of fandom [5, 6]. Culture jamming is mostly driven by the bottom-up momentum from the audience,

A. G. Ho (Ed.): AHFE 2019, AISC 974, pp. 95–105, 2020.
https://doi.org/10.1007/978-3-030-20500-3_10

which is highly compatible with the viral effect of users-generated content (UGC) via the networks of social media. The rise of digital mobilization facilitates the social influence of culture jamming, and it is adapted by commercial practitioners for various marketing objectives such as brand and product promotion [5, 7]. Such campaign objectives make good use of the viral impact of culture jamming to uphold the capitalist logics and corporate brand, and paradoxically go against the ethos of social movement and cultural resistance.

Shifting objectives and practices of culture jamming unleash a series of research questions. Firstly, to what extent the rhetoric strategy of culture jamming could work for the mainstream values, such as promotion of corporate brand and authorities? If so, under what conditions culture jamming could derail from its original theme of social resistance? These questions echo with the growing social attention to viral marketing in networks of social media. They also shed lights on how campaign designers and operators could adopt culture jamming as a mean of promotion for their products and clients.

In this paper, we will discuss the above questions by explicating a campaign launched by the Fire Services Department (FSD) of Hong Kong government in November 2018, in which a fictitious icon called "Anyone" (In Chinese: 任何仁, literally means anyone) was announced as the "spokesperson" to promote life-saving skills and initiatives. The icon defies the masculinist image of firemen, and humorously performs the life-saving skills with playful lyrics of pop music. Despite the initial controversies, this campaign has triggered roaring viral responses from the social media and salient coverage of mainstream media. More interestingly, while the figure of "Anyone" is imitated by various stakeholders for their own promotion, the main message of the FSD has remained intact, and the departmental image of FSD is also positively perceived by the public. The campaign is illustrative of how the rhetoric strategy of culture jamming is effectively employed by social authorities to promote institutional message and garner positive image. It is impetus to the scholarly dialogue of culture jamming, its marketing effectiveness and the strategies to employ culture jamming for corporate promotions.

2 Culture Jamming: Social Resistance and Beyond

Scholarly deliberation of culture jamming has been surrounding the conceptualization and practices of socio-cultural resistance. In the first stance, culture jamming refers to cultural practices which aim to derail the hegemonic discourses of capitalistic logic, mega-narration rationalizing social authorities and the "common sense" which is taken for grant [1, 4]. The term jamming implies disruption and derailment to the mainstream cultural meaning (re)produced in societies. Originated in the thesis of social movement, writings of culture jamming reiterate the sensitivity to cultural hegemony embedded in the unchallenged "common sense": corporate brands which encourage over-consumption, social imaginary of rules and order which make obedience to authorities as an unnoticed habit in everyday life, and the rituals and rites which justify social hierarchies and inequalities [3]. Cultural disruption and derailment of hegemonic ideologies could effectively alert the people about the alternative social imaginary vis-à-vis those depicted by the "common sense". Parodies, satires and mockeries against well-known brands, icons and authorities, for examples, could foster people's cynical perception of their familiar capitalist brands and practices [4]. Such "cultural hijacking" is an effective mean for the formation of resistant community through everyday life practices [8].

Derailment and disruption to hegemonic ideologies unravels alternative social imaginary of subaltern public spheres [9]. Practices of culture jamming often associate with socio-cultural advocacy of marginalized discourses and sub-culture, such as class inequalities, racial discrimination, labour exploitation, LGBTQ and alternative lifestyle [3, 10]. After all, the conceptual underpinnings of culture jamming are defiant against the mainstream social values. Culture jamming is therefore in line with the fostering of subaltern discourses. In addition, culture jamming also propels bottom-up participation from the people [11]. Disrupting and "hijacking" hegemonic discourses and social mainstream values entail creative inputs from the people for generating new social meaning, and the massive circulation of the "new" alternative cultural readings [2]. The ethos of culture jamming hence is highly compatible with the viral culture of social media networks, which characterizes with decentralization of controlling power, deinstitutionalization of organizational structure, and deprofessionalization of cultural participation and editorial process [12]. Culture jamming is therefore facilitated by the rise of social media and its viral culture.

While stemming from the ethos of socio-cultural resistance, some scholars explicate the possible derailment of culture jamming from the theme of social movement. An interesting argument is whether hegemonic formation and cultural domination are still viable in the age of digital culture and fragmented discourses [5]. As social media networks facilitate the creative inputs of user-generated content (UGC) by non-professionals, organizational gate-keeping of social information becomes less viable [12], presence of authoritative and hegemonic discourses *per se* in digital age hence is a polemic question. Furthermore, in the digital age of fragmented discourses, culture jamming ironically becomes a mean for consolidating fandom to corporate brand by rallying own fans to attack the rival brands [13]. Such 'hate-strategy' takes advantage of the creative inputs of culture jamming for supporting and strengthening the capitalistic logic of corporate branding.

Despite the rise of digital age and the proliferation of UGC, media scholars remind us that the social influence of mass media in agenda-setting should not be underestimated. Amid the over-supply of social information in digital era [14], credibility of mass media and news is still vital to the perception of social facts in the mind of people in general [15]. While online/social media could facilitate the rise of social sentiment, mass media to a certain extent could define the social facts perceived by the general public [16]. In addition, news professionals will also actively incorporate the UGC from social media into their professional paradigm instead of being passively moved by the sentiment manifested in social media [17].

In brief, while originating from the ethos of social movement for disrupting and "hijacking" the mainstream values, culture jamming is vulnerable to the instrumental goals of social authorities and corporate brands. Firstly, social authorities and corporates are aware of the viral power of culture jamming and are willing to make effective use of it. Secondly, in the digital age of discursive fragmentation and informational oversupply, credibility of news discourses and mass media is still a possible leverage of influence over the viral momentum of social media networks. In the following writings, we will illustrate how the FSD of Hong Kong government, as a respectful authority, makes an effective leverage of culture jamming to promote its message and image across both mainstream and social media by the creative campaign of "Anyone" (任何仁).

3 The Campaign of "Anyone"

The campaign of "Anyone" was launched by the FSD of Hong Kong government in early November 2018. "Anyone" is a fictitious faceless, human-shape blue character. As indicated by the scripts in the campaign videos[1], the campaign aims to educate the public of simple life-saving skills and encourage all people to safe those who are in urgent need immediately. The Chinese pronunciation of the fictitious figure "Anyone (任何仁)" literally refers to "anyone", which conveys the message that all people, no matter receiving professional training or not, could and should save those who are in critical health conditions. Despite the featuring of actress and sport star, the blue, faceless human figure of "Anyone" still captured most social attention. Firstly, on the contrary to the masculinist outlook of firemen, "Anyone" demonstrates a plump body shape with a fatty belly, and playfully sings and dances while demonstrating the simple life-saving skills. Depicting firemen as such blue, faceless and plump men triggered controversies accusing that whether the campaign has denigrated the image of firemen [18]. Representatives from the FSD said the character of "Anyone" was deliberately constructed as a layman instead of highlighting the masculinity and physical strength of firemen, as the campaign emphasizes that all people—no matter well-trained or not—could and should offer a helping hand to save those who are in critical health needs [19]. Secondly, the outlook of "Anyone" was soon identified by the public as a character which appeared in an episode of Japanese adult-video—"Oh, Invisible Man" (透明人間)—which depicted how a faceless invisible man sexually harass women [20]. Public responses to the campaign of "Anyone" conveyed the elements of culture jamming. The faceless and plump human figure has disrupted the masculine image and strong character of firemen. The cultural association of a pornographic figure in Japanese adult-video is obviously not in line with the solemn image of firemen.

Such controversial campaign of FDS, however, receives roaring responses in both mainstream and social media in Hong Kong. Figure 1 delineates the keyword search results from "Google Trends" which relate to "Anyone". "Google Trends" is an analytical tool illustrating the relative frequency of searching times of designated keywords. It does not reveal the actual search numbers. Instead, it informs us the relative percentage of searching frequency with respect to the highest searching number throughout the assigned period. We input four Chinese keywords: "Anyone" (任何仁), "Anyone" (任何人), firefighting (消防) and "Oh, Invisible Man" (透明人間). The first

[1] Please refer to the links in Youtube for related videos: https://www.youtube.com/watch?v=P9mWj-QMc10 (the video launched by the Facebook account of FSD on 4 Nov 2018, and https://www.youtube.com/watch?v=vKfVmNCpuZM (the video launched by the Facebook account of FSD on 8 Nov 2018).

two keywords are the same with each other in terms of Chinese pronunciation ("Ren He Ren" in terms of *Pinyin* Chinese), but they are different in the third Chinese character ("仁" VS "人", both are read as "Ren"). The first character (任何仁) is the original name designated by the FDS, while the second one (任何人) is the typo which means "human-beings". As the second Chinese term (任何人) is highly compatible with the key message of "anyone can save people", it attracts a high searching frequency in Google search engine too. The keyword of "firefighting" (消防, the colloquial depiction of the FDS in Hong Kong) is included, so that we can observe whether people's interest in the FDS relates to the campaign of "Anyone". Lastly, the keyword "Oh! Invisible Man" (透明人間) could show the association of the campaign and the cultural imagination of this Japanese adult-video, which is illustrative to the power of culture jamming.

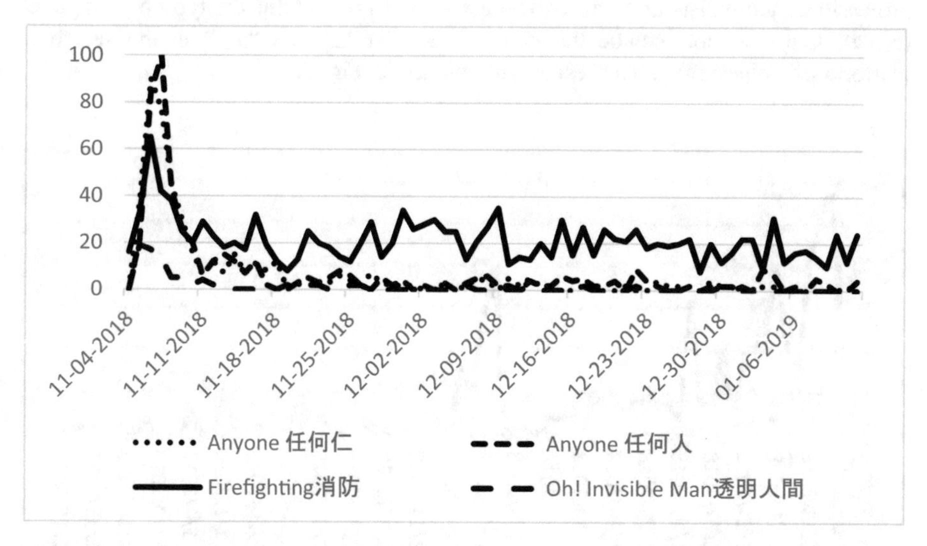

Fig. 1. Google searching frequencies of keywords relating to the campaign of "Anyone". The numbers in the y-axis represent the relative percentage to the highest frequency of the dataset, and the x-axis shows the dates in the format of MM-DD-YYYY

Figure 1 indicates that Hong Kong people were keen to search for the campaign of "Anyone" in the first few days after the FDS launched this fictitious figure on 4 November 2018. The lines which represent the searching frequencies of "Anyone" (任何仁) and "Anyone" (任何人) respectively, quickly climbed up to the peak on 6 November 2018 and 7 November 2018. They dropped down rapidly in the few days afterward. The trend shows that the campaign of "Anyone" successfully seized the social attention in the first week after its launching. Moreover, the searching frequency of the keyword "firefighting" (消防) also demonstrated a corresponding increase and decrease with the trends of both Chinese keywords of "Anyone". As "firefighting" and the FDS are significant to society, it maintained a fluctuating presence when both Chinese terms of "Anyone" dropped to a flatting low searching frequency. On the contrary, the searching frequency of the keyword of "Oh! Invisible Man" (透明人間) briefly increased from 4 November to 5 November 2018, and then disappeared gradually to a very low level. The results of search frequency illustrate that the campaign of "Anyone" has successfully seized the attention of Hong Kong people when it was launched, and the social attention to "Anyone" has also drawn people's interest in the work of the FDS. Conversely, while the cultural association of "Anyone" to Japanese adult-video made some noise initially, it was apparently overwhelmed by people's enthusiastic interest in "Anyone" and "firefighting" since the very beginning of the campaign.

In addition to the searching frequencies of related keywords in Google, we also examine the number of news coverage in mainstream media secured by the campaign of "Anyone". We input the keyword "Anyone" (任何仁) in the searching engine of "Wisenews"—a database collecting all news coverage of Hong Kong since 1998. Since professional journalists and news coverage should not commit the typo of "Anyone" (任何仁), we do not include the typo of "Ren" ("仁" VS "人") in the searching platform of "Wisenews". The results are enlisted in Fig. 2.

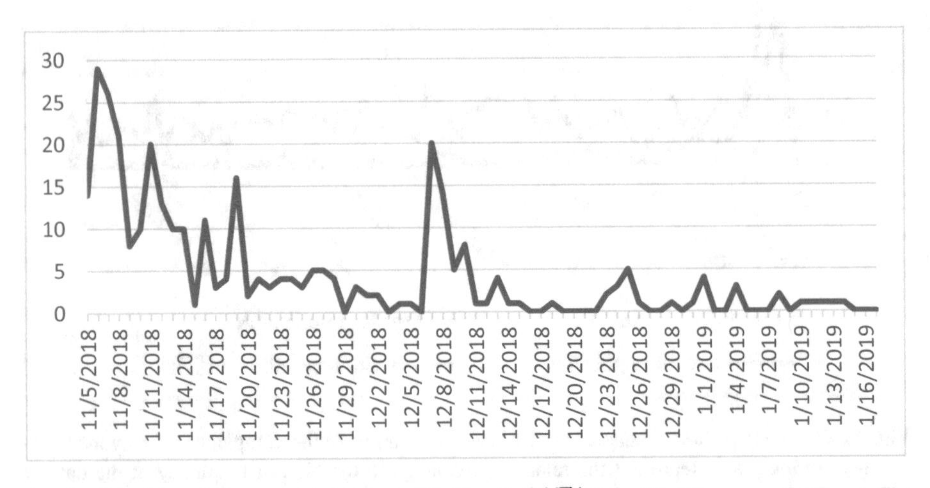

Fig. 2. Searching frequency of the term "Anyone" (任何仁) in the Wisenews database. The numbers in the y-axis represent the number of news items, and the x-axis shows the dates in the format of MM-DD-YYYY

Figure 2 shows that the news coverage of "Anyone" surged rapidly once the campaign was launched, and gradually fluctuated in a downward trend afterward, except the short upsurge in early December 2018. The pattern is compatible to those frequencies of keyword search in Google. The brief increase of news coverage in early December 2018 was most likely due to the media interviews with the production team of the campaign of "Anyone" and its head [21]. Juxtaposition of Figs. 1 and 2 indicates an intriguing interplay between news coverage and the online world. Upsurge in Google searching frequencies of related keywords represents people's interest via interactive and mobile platforms. Salience of news coverage suggests journalistic agenda set by media organizations. In fact, the FSD uploaded the campaign videos on its Facebook account on 4 November and 8 November 2018 and held press briefings in between to explain the theme and objectives of the campaign. A significant "gimmick" in the first week of the campaign is that there were rumors claiming that the FSD would relinquish "Anyone" from the campaign owing to its playful manner, which triggered fierce objection from networks of social media—and the FSD eventually denounced the rumor and announced that "Anyone" will continue to be the "spokesperson" of the department in its press conference [22]. The interplay between news coverage and online sentiment suggests that while the viral effect of digital platform is enormous, news media still enjoy a certain kind of "agenda-setting" influence [23].

Apart from the quantitative analyses of Google trends and news coverage, we also scrutinize the news discourses covering the campaign of "Anyone". Table 1 summarizes our textual analysis of related news discourses in according with the conceptual framework outlined by Gamson and Lasch [24]. It delineates three major news discourses covering the campaign of "Anyone". The first one is the theme message of the campaign itself: "live-saving". In the campaign video launched by the FSD on 8 November 2018 (please refer to the link listed in footnote 1), "Anyone" taught a boxing star in Hong Kong—Mr. Rex Tso (曹星如) a motto: "brave could save one's life" (敢就救到人), in order to encourage people who do not receive professional first-aid training also offer helping hand to those who are in critical medical needs. This motto becomes the catchphrase of news coverage about the campaign [19]. In December 2018, the campaign production team leader who made "Anyone" explained his creative idea in press coverage. He said that the idea stems from the theme that saving one's life could save a family [21]. Such campaign messages became the catchphrases appeared in the press. Moreover, the FSD also arranged "Anyone" to demonstrate life-saving skills in press conference and other educational promotion to earn news coverage [19]. As a result, professional bodies, lawmakers and the public called for the establishment of Good Samaritan Law to protect people from unintended legal liability when saving others [25], and there was editorial urging the education of life-saving skills in schools [26]. Lastly, the news discourse of "life-saving" also appreciated the liberal attitude of the FSD to such innovative campaign [27].

Table 1. Major news discourses covering the campaign of "Anyone"

News discourses	Life-saving	Public relations skills of the FSD	Political attack
Metaphor/catchphrase	Brave could save one's life (in Chinese) Saving one life is saving a family (in Chinese)	Overwhelming responses on Facebook (in Chinese) Colloquial metaphors of the popularity of "Anyone": "Siphoning Water" (in Chinese), "Explosion" (in Chinese)	Democrat politician Lee Cheuk Yan imitates "Anyone" in order to "siphons water" (in Chinese)—taking advantages of the popularity of "Anyone"

(*continued*)

Table 1. (*continued*)

News discourses	Life-saving	Public relations skills of the FSD	Political attack
Exemplar/depiction	"Anyone" demonstrate life-saving skills	"Anyone" is busy (which means appear everywhere)	Lee Cheuk Yan "consumes" the popularity of the FSD
Root/consequences	The public urges setting up Good Samaritan Law Promoting life-saving skills in schools	Other governmental departments, public authorities, and commercial organizations make icons which imitate the outlook of "Anyone"	Lee Cheuk Yan does not have credibility
Principle/moral appeal	Liberal attitude of the FSD	"Anyone" is a highly successful campaign	The FSD said "Anyone" does not have any deals with Lee Cheuk Yan

The second major news discourse surrounding the campaign of "Anyone" is the appreciation of its public relations skills. Headlines of press coverage indicate metaphors and catchphrases of "overwhelming responses on Facebook" (FB 洗版), "siphoning water" (抽水) and "explosion" (爆紅、熱爆). Such colloquial terms indicate the popularity of "Anyone" in the social media networks. News coverage depicts that "Anyone" is busy, which means the image of this blue and faceless human-shaped figure appear everywhere in Hong Kong. Gradually, other governmental departments, commercial organizations and non-governmental organizations imitate the image of "Anyone" in their own promotional campaigns: Environment Bureau, Hong Kong Red Cross, and beer-seller, for examples [28]. Copycat from other organizations illustrates a roaring social impact of the FSD's campaign.

Lastly, the pro-China press attacked a democrat politician (political rivalry to pro-China camp in Hong Kong)—Mr. Lee Cheuk Yan when Lee imitated "Anyone" in his political canvassing. The pro-China press used the term "siphoning water" (抽水) in a negative connotation of "taking advantage of something" and depicted Lee as a cunning politician who "consumed" the popularity of the FSD and "Anyone" [29]. They cited the response from the FSD that the department did not have any deals of cooperation with Lee of using "Anyone" for political canvassing and attacked Lee's personal credibility [30].

4 Discussion and Conclusion

The campaign story of "Anyone" explicates numerous lessons to making use of culture jamming for promotional purposes. Firstly, while the viral impact of social media is widely recognized, the social influence of mainstream media is still significant.

The FSD announced the key message of the campaign of "Anyone" via press conferences and news releases in addition to posting the campaign videos on its Facebook account. The department also denounced the rumor of relinquishing "Anyone" via the platform of news discourse. Mainstream media and news discourses, to a certain extent, still enjoy a higher level of social trust over the interpersonal gossips transmitting in networks of social media. Interplay between news releases and online sentiment enables the FSD to send out its intact official message amid its interaction with the public via platforms of social media.

More importantly, the success of the campaign also rides on the social trust to the FSD and firefighting services in Hong Kong. Firstly, despite the initial controversies over the image of "Anyone" and its association with Japanese pornographic adult-video, Google searching results show that the public soon focused back on the fire services. Analyses of news discourses further reveal the massive circulation of the key message of "life-saving" and public appreciation of the public relations skills of the FSD. The positive response shown in Google trend and active follow-ups of news media are illustrative of the credibility of the FSD in Hong Kong. Conversely, when democrat politician Lee Cheuk Yan attempted to copy the gimmick of "Anyone" in his own political canvassing, the pro-China press reacted by criticism and cynical remarks. Despite pro-China press could not represent the general public opinion of Hong Kong, hostile news discourse against Lee Cheuk Yan demonstrates that campaign of culture jamming should be taken by credible institution or individuals, or otherwise the campaign is highly vulnerable to "back-fire" from the public or other social stakeholders.

A final remark to the campaign of "Anyone" is that the story of culture jamming could cut across the fields of social movement and business branding. Given appropriate conditions—such as interplay with mainstream media and the institutional and individual credibility incurred in the campaign—culture jamming could possibly be an effective mean for brand promotion on behalf of social authorities. This case study, although is far from a representative analysis of varying sorts of creative campaigns, pinpoints the gradual blurring line between social movement and business branding when we examine the motives and effectiveness of culture jamming. With the growing awareness to the viral effectiveness of culture jamming by social authorities and business corporations, adaptation of culture jamming in the promotion of mainstream values and corporate image is a salient topic for both scholarly and professional dialogue.

References

1. Anderson, J., Kincaid, A.D.: Media subservience and satirical subversiveness: The Daily Show, The Colbert Report, the propaganda model and the paradox of parody. Crit. Stud. Media Commun. **30**(3), 171–188 (2013)
2. Bruun, H.: Political satire in Danish television: reinventing a tradition. Pop. Commun. **10** (1/2), 158–169 (2012)
3. Verson, J.: Why we need cultural activism. In: Collective, T. (ed.) Do It Yourself: A Handbook for Changing the World. Pluto Press, London (2007)
4. Harold, C.: Pranking rhetoric: "culture jamming" as media activism. Crit. Stud. Media Commun. **21**(3), 189–211 (2004)

5. Cammaerts, B.: Jamming the political: beyond counter-hegemonic practices. Continuum: J. Media Cult. Stud. **21**(1), 71–90 (2007)
6. Chatterton, P., Pickerill, J.: Everyday activism and transitions towards post-capitalist worlds. Trans. Inst. Brit. Geogr. **35**, 475–490 (2010)
7. Soar, M.: The first things first manifesto and the politics of culture jamming: towards a cultural economy of graphic design and advertising. Cult. Stud. **16**(4), 570–592 (2002)
8. Boukes, M., Boomgaarden, H.G., Moorman, M., de Vreese, C.H.: At odds: laughing and thinking? The appreciation, processing, and persuasiveness of political satire. J. Commun. **65**(5), 721–744 (2015)
9. Ibrahim, A., Eltantawy, N.: Egypt's Jon Stewart: humorous political satire and serious culture jamming. Int. J. Commun. **11**, 2806–2824 (2017)
10. Warner, J.: Political culture jamming: The dissident humor of "The Daily Show with Jon Stewart". Pop. Commun. **5**(1), 17–36 (2007)
11. Landreville, K.D., LaMarre, H.L.: Working through political entertainment: how negative emotion and narrative engagement encourage political discussion intent in young Americans. Commun. Q. **59**(2), 200–220 (2011)
12. Atton, C.: Alternative and citizen journalism. In: Wahl-Jorgensen, K., Hanitzsch, T. (eds.) The Handbook of Journalism Studies, pp. 265–278. Routledge, NY (2009)
13. Luedicke, M.K., Giesler, M.: Contested consumption in everyday life. In: Lee, A.Y., Soman, D. (eds.) Advances in Consumer Research, vol. 35, pp. 812–813. Association for Consumer Research, Duluth (2008)
14. Bennett, W.L., Iyengar, S.: A new era of minimal effects? The changing foundations of political communication. J. Commun. **58**(4), 707–731 (2008)
15. Schudson, M.: Why Democracies Need an Lovable Press?. Polity Press, Cambridge (2008)
16. Zhou, Y., Moy, P.: Parsing framing processes: the interplay between online public opinion and media coverage. J. Commun. **57**(1), 79–98 (2007)
17. Lee, F.L.F.: News from YouTube: professional incorporation in Hong Kong newspaper coverage of online videos. Asian J. Commun. **22**(1), 1–18 (2012)
18. Hong Kong Economic Journal: Outlook of "Anyone" of fire services department is mocked by netizens. Hong Kong Economic Journal, A14 (2018, in Chinese)
19. Ming Pao Daily News: Fire services department teaches CPR, many imitate the popular blue human figure called "Anyone". Ming Pao Daily News, A12 (2018, in Chinese)
20. Oriental Daily News: Fire services department promotes "Anyone", whom is mocked as a character of Japanese adult-video. Oriental Daily News, A10 (2018, in Chinese)
21. Ta Kung Pao: Saving one's life is saving a family, said the "father" of "Anyone". Tak Kung Pao, A07 (2018, in Chinese)
22. Ming Pao Daily News: "Anyone" does not disappear and claims his own safety. Ming Pao Daily News, A17 (2018, in Chinese)
23. Lee. F.L.F.: Golden 48 hours of online searching: how fire services department plays with "Anyone". Ming Pao Daily News, P02 (2018, in Chinese)
24. Gamson, W.A., Lasch, K.E.: The political culture of social welfare policy. In: Spiro, S.E., Yuchtman-Yaar, E. (eds.) Evaluating the Welfare State: Social and Political Perspectives. Academic Press, London (1983)
25. Headline Daily: People are afraid of the legal liability owing to saving others, the medical sector urges the establishment of Good Samaritan Law to erase concern. Headline Daily, P12 (2018, in Chinese)
26. Sing Pao: Mandatory life-saving education in schools is better than the Good Samaritan Law. Sing Pao, Editorial, A03 (2018, in Chinese)
27. AM730: Rumors of no more real-man show, "Anyone": my boss is liberal. AM730, A26 (2018, in Chinese)

28. Sing Pao: Governmental departments and business brands all-out amid rising popularity of "Anyone"—the whole city keeps making good use of it. Sing Pao, A04 (2018, in Chinese)
29. Ta Kung Pao: Shameless! Lee Cheuk Yan repeatedly "consumes" the firefighers. Ta Kung Pao, A10 (2018, in Chinese)
30. Wen Wei Po: Fire Services Department: no cooperation with "Anyone". Wen Wei Po, A13, 16 November 2018

Digital Generation's Meaning-Making in Web-Based Communication Activities

Pui Wa Chau(✉)

Toppswill, Kwun Tong, Hong Kong
r.ruthchau@gmail.com

Abstract. In the communication process, the human is not the message recipients solely but also the interrupters of the message meaning. Theorists of social studies developed some concepts of message meaning as semiotic theories. These theories highlighted the essential features of individual. Introducing the approaches of interpreting semiotics concepts, the meaning-making process was considered according to its meanings, which certainly involved an interpreting agent. These theories were the principles to explore various sign system with different cultural backgrounds. However, the limited understanding was explored for the digital generation who have grown up with complicated digital information system and advanced communication technologies. This paper offers a brief revision of the semiotic theories. It also examined the roles of interpreting meanings those involved in the semiotic process through web-based communication activities. Content analysis of several web-based communication activities would be employed as the research method in this study. It is expected that the result of this study would provide insight for more exploration of the communication principles of the digital generation.

Keywords: Meaning-making · Web-based communication · Digital generation · Communication system

1 Introduction

Various issues, as well as challenges, were discussed in the communication design field. Influenced by contemporary media technologies, the communication process was analysed from two aspects from production as well as the reception. The discussions about production mainly considered the exchanges among various signals. The-se discussions contributed the concepts about how the messages and their performance for creating influence on the audience. However, the limited understanding was explored for the digital generation who have grown up with complicated digital information system and advanced communication technologies. This paper offers a brief revision of the semiotic theories. It also examined the roles of interpreting meanings those involved in the semiotic process through web-based communication active-ties. Content analysis of several web-based communication activities would be employed as the tool in this research study. It is expected that the result of this study would provide insight for more exploration of the communication principles of the digital generation.

A. G. Ho (Ed.): AHFE 2019, AISC 974, pp. 106–113, 2020.
https://doi.org/10.1007/978-3-030-20500-3_11

2 Meaning-Making in Web-Based Communication Activities

In the communication process, the human is not the message recipients solely but also the interrupters of the message meaning. Some research studies were conducted for understanding the factor of constructing an audience's trust, satisfaction as well as loyalty through web-based communication. This paper aimed to focus on drivers of customers revisiting a web-based vendor. The current studies on web-based communication indicated a significant preference for the audience. Some website investigated how to obtain an audience's trust and satisfaction. Nishino et al. [1] compared to the application of the same communication procedures in five counties in parallel. The research team proposed that they were the essential factors for building up the audience's loyalty based on the audience's data collected in his study. Scholars discovered that the feature of effective message delivery mostly included semiotic. O'Halloran and his time explored that designers created messages based on how the audience receives the material such as words and images during the communication process. The way of composing messages and its potential for making meaning were concerned in the research field of social semiotic studies [2, 3].

3 Semiotic Theories

Took reference from the perspectives of social studies, some concepts of message meaning was developed based on semiotic theories. According to previous semiotic studies referred to the interactive activities among words and images. They were the essential elements for visual storytelling [4]. The effectiveness of how semiotic theories application in a society depends on the social impact of communication as well as the rhetoric of criticism in communication. In order words, these semiotic theories highlighted the essential principles of communication. After introducing the concepts of interpreting into semiotic theories, the process of meaning-making was considered a condition of signification. In other words, these theories were the principles to explore various sign system with different cultural backgrounds. Holsanova and his research team [3] combined the concept of semiotic as well as communication process from cognitive approaches. They presented new methods to derive the changes of visual fragmentation those influenced by the underlying cognitive thinking processes. By applying the interdisciplinary methods and technologies such as eye-movement tracking as well as verbal agreement checking, visual fragmentation was traced as the initial actions of the audience individually. Based on this point of view from social semiotic theories, the audience thus perceived images and words, and then, made meaning. The research result of Holsanova 's study provided an alternative point of view on investigating how images and words would be fragmented and interpreted by the audience. Some studies investigated the rhetorical tradition view [5]. However, there were still have needs to develop methods those for explaining the semiotic acts through advanced analysis. Multimodel theories thus were developed and contributed. Multimodel theories reflected the close relationships between the audience's ability to select information and the process of gathering information. It explained the domain role of the audience who ultimately select the appropriate information. The audience selected the topics that they prefer to explore, the way they want to process as well as the way they interpret. In other words, multimodel theories illustrated the active roles of the audience in interactive communication. Ko et al. [6] investigated how the audience interacts with each other with visual. Their findings of how audience obtained message as well as situation context was similar to the concepts of the multimodal theories. Their studies also explained that different audience could perceive the same visual differently. They could also ascribe very different interactions with the messages they obtained. Therefore, it was essential to understand how specific audience interacts with messages through the multimedia and create meaning. Based on the multimodal theories, the visual perception was determined by appeal elements included form, texture, colour, motion as well as top-down elements referred audience' features included personal interests, preferences, knowledge as well as cultural background. Therefore, the approach of perceiving adopted by audience influenced their understanding and memory through the content of the multimedia which was guided partly by visual and partly by semantic elements. The question of how the audience evaluating the semantic elements have not yet investigated by the researchers in visual communication studies.

4 Semiotic Process in Web-Based Communication Activities

The role of a semiotic visual with the semiotic process in web-based communication activities was investigated. In order to prevent consuming too long time as well as too much energy in applying the relative traditional approaches for obtaining information through interviews, servers as well as focus groups, researchers are now able to download information and raw data from the web-based activities without direct communication with the audience. Although this case might not be able to be applied in all kinds of research fields, most research studies in social sciences, information communication nowadays paid more concerns about their participants' feedback on their attitudes, personal preferences, as well as their behaviours. Some scholars proposed that the audience obtained benefits from the web-based communication [7]. In the past study, the data from web-based communication presented a variety of structures, and some of them were not organised well; content analysis lends itself as the most appropriate method to analyse such data. This result examined the pervious concept developed for investigating the approach to create scientific methods for ensuring the quality of the contents of messages [8]. He proposed that content analysis would be one of the possible methods for comparison. Based on the investigation by Krippendorff, several advantages of applying content analysis for comparing personal interests, preferences, knowledge as well as cultural background were identified in pervious social sciences studies. Krippendorff pointed out that although content analysis might be not obtrusive and structured thoroughly. It was sensitively and was coped with a large amount of information. Krippendorff examined the effectiveness of content in his study. He found that the words and images in the communication process with different media those did not design individually would be less effective.

After understanding the challenge of web-based communication, some studies were conducted for ensuring the quality of the contents of messages through qualitative measurement. Some scholars investigated how cultural values were applied in web-based communication by studying their content [9]. The study of Singh and Baack explored the influence of cultural differences on the organisation of web-based communication. The research team found there were similarities in the web-based communication with a model of cultural dimensions [10]. At the same time, Kim and Kuljis investigated the methods of manipulating content analysis in a research study to web-based material. They explained that the content those delivered by Web 2.0 technology have both distinct advantages and limitations. In the field of research, there were two types of data measurement generally: quantitative and qualitative measurement. Researcher sometimes applied both and sometimes only applied each. Most studies applied the quantitative measurements such as click rates, the length of surfing time on the website, and so forth. The sign of web communication should be examined by qualitative measurement, for example, trust, satisfaction and loyalty in international markets.

5 Research Method

Content analysis of several web-based communication activities was employed as the research method in this study. Some scholars investigated some approaches of analysing as the application of scientific methods to the documentary. Krippendorff obtained similar concepts with the priors, and then he defined content analysis as a research tool for creating reliable as well as valid reasoning from raw data to the related information. The research tool, content analysis enabled data to be analysed and restructured for applying in both qualitative and quantitative data. Qualitative content analysis was similar to textual analysis based on a participants' point of view typically. According to Krippendorff, the feedback from the participants was primarily interpretive information. Most of them were unable to adopt statistics for analysis. On the other hand, quantitative content analysis was a research tool for providing measurable as well as reliable information.

6 Research Process

Twenty participants were invited to be the research participants. They were to be divided into four groups. Each group were grouped by five participants. All participants were Year 2 undergraduate students. For analysing the content, a scientific method for variance analysis was adopted. It was a tool for analysing. The statistics in the analysis provided aggregate variability. This variability found among raw data would be divided into two factors systematically as 'systematic factors' as well as 'random factors'. The 'systematic factors' factors offered statistical influence on the found raw data while the 'random factors' did not. In this research study, the test of variance analysis was applied for determining the influence that individual variables provided on the related variable. The coefficient test of variance analysis was calculated through the formula of 'Mean sum of squares due to treatment' divided by 'Mean sum of squares due to error' [11]. This tool for variance analysis was designed for analysing factors that initially influenced the obtained data and information. While the analysis was finished, researchers would perform follow-up testing on the 'random factors' which contributed inconsistency to the obtained findings. The test of variance analysis was applied in this study. The test results provided additional information that aligned with the proposed regression models. The analysis of variance test allowed a comparison of more than two groups at the same time to determine whether a relationship exists between them. The result of the analysis of variance testing formula provided analysis of different groups of raw data for determining the variability among samples. If there were no real difference existed among the participants, which was called the null hypothesis, the result of the variance test analysing formula would be closed to 1. Fluctuations in its sampling mostly reflected the distribution. This reflected a group of distribution functions in an accurate way. Hence, the variance test analysing formula reflected the numerous degrees of freedom as well as the denominator degrees of freedom. Comparing to other techniques included questionnaire surveys, focus group, experiment tests, the contributions of content analysis attracted researchers with the least biased Another benefit. The procedure for undertaking content analysis was relatively simple. Therefore, content analysis, in particular, was an appropriated research method for analysing Web-based content. The content generated by participants would be reliable information for understanding participants without much interactions with them. The vast quantities of data could be investigated. It would be considered as an advantage employed for examining trends as well as the communication pattern of web-based content (Fig. 1).

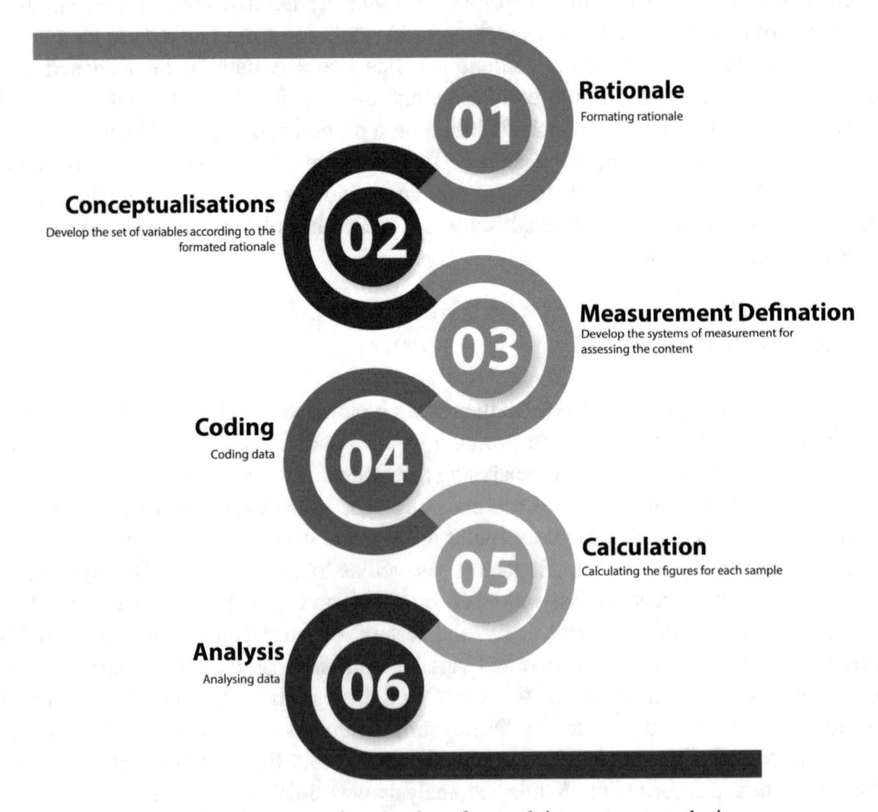

Fig. 1. The research procedure for applying content analysis.

7 Research Result

In this study, wide arrange of the variable as well as visual design elements those applied in web-based communication activities were investigated. The self-disclosed feedbacks of the participants were sufficient to illustrate how participants apply the variable as well as visual design elements in the content of their communication through their blogs. According to the comparisons of the features of the variable as well as visual design elements related to self-disclosure, it was found that the different dimensions of cultures would influence the posted content in the blogs. The raw data were obtained according to the application of the variance analysing. The results indicated that Hong Kong bloggers were more willing to emphases information about themselves (8.64) than Taiwan bloggers (6.88), yet this was not a significant disparity. The frequency of the occurrence of Taiwan blogger occupation was significantly higher than the case of Hong Kong blogs (10.40). Taiwan blogs more often revealed age (5.76) and more Taiwan bloggers provided a contact link to the author than the Hong Kong blogs (3.45). Therefore, the result indicated that the culture might not have that many impacts regarding the self-disclosure. However, when the study was further conducted, there was a big difference in the number of blogs registered for each country. Taiwan blogs were in the minority, and thus they were not representative enough. Therefore, another sample for Taiwan blogs was selected. However, the results of this part of the study were not yet complete at the time of this study. To summarise, the finding performed content analysis test by following the proposed procedures. The approach of content analysis proposed by Nuendorf was accessible and useful. However, his framework did not provide any detail guidance on the way for conducting sampling of content and to how to prevent the influence of the nature of the web-based communication.

8 Conclusion

This study expanded the knowledge about the impact of web site design for enhancing the audience's trust, satisfaction and loyalty. A human does not only take part as the message recipients solely in the communication process; the human is not the message recipients solely but also the interrupters of the message meaning. Theorists of social studies developed some concepts of message meaning as semiotic theories. These theories highlighted the essential features of an individual. Introducing the approaches of interpreting semiotics concepts, the meaning-making process was considered according to its meanings, which indeed involved an interpreting agent. These theories were the principles to explore various sign system with different cultural backgrounds. However, the limited understanding was explored for the digital generation who have grown up with complicated digital information system and advanced communication technologies. This paper offers a brief revision of the semiotic theories. It also examined the roles of interpreting meanings those involved in the semiotic process through web-based communication activities. Content analysis of several web-based communication activities would be employed as the research method in this study. It is expected that the result of this study would provide insight for more exploration of the communication principles of the digital generation. And also, the interrupters of the message meaning was suggested. Theorists of social studies developed some concepts of message meaning as semiotic theories. These theories highlighted the essential features of individual. Introducing the approaches of interpreting semiotics concepts, the meaning-making process was considered according to its meanings, which certainly involved an interpreting agent. These theories were the principles to explore various sign system with different cultural backgrounds. However, the limited understanding was explored for the digital generation who have grown up with complicated digital information system and advanced communication technologies. This paper offers a brief revision of the semiotic theories. It also examined the roles of interpreting meanings those involved in the semiotic process through web-based communication activities. Content analysis of several web-based communication activities would be employed as the research method in this study. It is expected that the result of this study would provide insight for more exploration of the communication principles of the digital generation. Designers who explored approaches for enhancing the market communication with web-based content those would implicate the research result. Limitations of the study were that both the local and foreign web sites applied were web-based communication by using one browser only and only a single task was applied. In the foreseeable future, further, investigate would provide broader covering topics such as mobile learning, learning technologies as well as individual learning networks.

References

1. Nishino, T., Hirata, O., Suzuki, S., Hara, M., Tsuchida, S., Ogata, M.: U.S: Patent No. 5,001,744. Washington, DC: U.S. Patent and Trademark Office (1991)
2. Kress, G.R., Van Leeuwen, T.: Reading images: The Grammar of Visual Design. Psychology Press, Hove (1996)
3. Holsanova, J., Rahm, H., Holmqvist, K.: Entry points and reading paths on newspaper spreads: comparing a semiotic analysis with eye-tracking measurements. Vis. Commun. **5** (1), 65–93 (2006)
4. Baisch, B., Raffa, D., Jung, U., Magnussen, O.M., Nicolas, C., Lacour, J., Herges, R.: Mounting freestanding molecular functions onto surfaces: The platform approach. J. Am. Chem. Soc. **131**(2), 442–443 (2008)
5. Bateman, J.: Multimodality and Genre: A Foundation for the Systematic Analysis of Multimodal Documents. Springer, Heidelberg (2008)
6. Ko, C.J., Braverman, I., Sidlow, R., Lowenstein, E.: CME article part 1: Visual perception, cognition and error in dermatologic diagnosis. J. Am. Acad. Dermatol. (2019)
7. Krippendorff, K.: Content Analysis: An Introduction to its Methodology. Sage Publications, London (1980)
8. Holsti, O.R.: Content Analysis for the Social Sciences and Humanities, Reading. Addison-Wesley Publishing, MA (1969)
9. Singh, N., Baack, D.W.: Web Site adaptation: a Cross-Cultural comparison of U.S. and Mexican web sites. J. Comput. Mediated Commun. **9**(4), 1 (2004)

10. Hofstede, G.: Cultures and Organisations: Software of the Mind: Intercultural Cooperation and its Importance for Survival. McGraw Hill, New York (1991)
11. Kim, I., Kuljis, J.: Applying content analysis to web-based content. J. Comput. Inf. Technol. **18**(4), 369–375 (2010)

Creative Arts

Improvement of Safety Sign Design Based on Analysis Comprehensibility Test Result

Chuan-yu Zou[1(✉)] and Yongquan Chen[2]

[1] LHFE Key Laboratory of Human Factors and Ergonomics (CNIS), Beijing, China
zouchy@cnis.gov.cn

[2] Research Centre of Way Guidance, China National Institute of Standardization, Beijing, China

Abstract. **Object**: To improve the effectiveness of safety signs, it is necessary to study their comprehensibility and design features. **Methods** 100 respondents and 6 safety signs were selected to conduct research on the comprehensibility and design features of safety signs. **Results** The design features of safety signs directly affect their comprehensibility. Abstract, and relevance can be taken care of during design, while familiarity can be improved by pre-education and training.

Keywords: Safety sign · Design feature · Comprehensibility test

1 Introduction

1.1 Background

Safety signs can "give a general safety message, obtained by a combination of a colour and geometric shape and which, by the audition of a graphical symbol, gives a particular safety message" [1]. Safety message helps people avoid potential hazards in their daily work and life. Falling objects injury may occur on the construction site, mechanical crushing may occur in the processing workshop, and vehicle collision may occur in transportation operations. In Albert et al. study, alarming workplace incident rates, especially in the construction sector, continue to be of global concern [2], while potential hazards in daily life are more diverse. Small parts on toys may be eaten by children, electrical leakage may cause electric shock, and stepping on the stairs may cause a fall. In Zhang Dianye et al. research, the causes of potential dangers of loss of personal property can be divided into two categories: one is the potential danger itself; another one comes down to human negligence [3].

A safety sign is a sign that warns the public and staff about the dangerous situation in public places, workplaces or the surrounding conditions and guides people to take necessary actions to avoid potentially injuries. Safety signs can alert people to potential hazards and prevent hazards to avoid accidents. When a hazard occurs, necessary step should be taken to remove potential hazards firstly. If the potential hazard cannot be removed, a safety sign should be used to instruct people to flee as quickly as possible, or instruct people to take correct, effective, and effective measures to contain the

A. G. Ho (Ed.): AHFE 2019, AISC 974, pp. 117–123, 2020.
https://doi.org/10.1007/978-3-030-20500-3_12

hazard. Not only should the type of safety sign match the content of the warning, it must be correctly comprehend, otherwise it is difficult to fully exert warning effect.

Safety signs with good design can help people quickly identify and stay away from potential danger [4].

1.2 Current Standards Status

There are 26 standards related to safety signs in China's national standards, including the basic standards for designing safety signs and the special standard safety signs for each field. Among them, GB/T 2893 series standard [5] and GB 2894 [6] are the basic standards for designing safety signs. National standards such as GB 13495 [7] and GB 16557 [8] regulate special safety signs in the fields of fire protection and sea-going ship.

There are 16 currently effective standards related to safety signs in International Organization for Standardization (ISO). Those standards also categorize into the basic standards for designing safety signs and the special safety signs standards for each field according to their own needs. ISO/TC 145/SC 2 "Safety identification, signs, shapes, symbols and colours", as a horizontal technical committee responsible for safety signs, has developed and published a number of basic safety sign standards, the most important of which is ISO 7010 [9]. The standard specifies 216 safety signs in five categories: "Signs indicating an evacuation route, the location of safety equipment or a safety facility, or a safety action", "Fire equipment signs", "Mandatory action signs", "Prohibition signs", and "Warning Signs". Other standards related to safety signs mainly specify safety signs in the workplace, distributed in areas such as Ships and marine technology, Powered industrial trucks, Cranes.

1.3 Current Research Status

Based on the safety science theory and the existing research results, combined with in-depth interviews and large-scale questionnaire surveys of 13 companies, Yuan [10] conducted an empirical study on the factors affecting the effectiveness of safety signs was conducted. From the perspective of ergonomics system optimization, Hu et al. [11] constructed the safety sign effectiveness evaluation index system by dividing the cognitive behavior stage of construction workers. The research shows that the effectiveness of safety signs is noticeable in the attention stage and the identification stage. The security tendency of comprehension and judgment stage and the compliance of the compliance stage of behavioral stage, among which the value of security tendency weight is the highest. Chan et al. [12] investigated the effects of prospective-user factors and five cognitive sign features on guessability of safety signs. Easterby [13] tested symbolically coded sign variants in the field using a structured national random sample of 4000 respondents. In this study, prior experience of the sign was found to enhance comprehension, increasing the correct comprehension rate by a factor of 1.5 ~ 2.0 times the correct comprehension rates obtained with those respondents who affirm that they have not seen the sign before.

1.4 Safety Signs

In ISO 7010 [9] five categories of safety signs are specifies: "Safe condition signs", "Fire equipment signs", "Mandatory action signs", "Prohibition signs", and "Warning Signs".

Safe condition signs are safety graphical symbols combined with green rectangle/square. Fire equipment signs are safety graphical symbols combined with red rectangle/square. Mandatory action signs are safety graphical symbols combined with blue circular area. Prohibition signs are safety graphical symbols combined with red circle with red slash. Warning Signs are safety graphical symbols combined with yellow triangle with black border. Those five categories safety signs and their specified colours and shapes are shown in Fig. 1.

Fig. 1. 5 categories of safety signs

2 Methods

2.1 Respondents

100 recruited respondents (50 males and 50 females) participated in the test. Their age ranged from 20 to 25 years old. All respondents had normal or corrected visual acuity and did not participate in a similar safety sign test questionnaire.

2.2 Safety Signs

Comprehensibility test [14] was conducted to evaluate safety signs. In order to reduce the impact of safety geometric shapes and numbers of safety signs on the comprehensibility of safety signs, 3 safe condition signs and 3 mandatory action signs in GB/T 31523 were chosen to take comprehensibility test. They are: "Break to obtain access" "General alarm" "Lifebuoy" "Wear a mask" "Use handrail" "Secure gas cylinders" (see Fig. 2).

Fig. 2. 6 safety signs in comprehensibility test

2.3 Testing Conditions

The test was carried out in a multimedia classroom using computers. All test materials are made into electronic files that are displayed on the screen page by page. The first page is the personal information page, the second page is the test description page, followed by the relevant test pages. Each test page has a home button, and respondents can view the test description page at any time. Each test page cannot be modified after be submitted.

2.4 Comprehensibility Test

Comprehensibility test is a method to evaluate the extent to which a safety sign is likely to be interpreted with the intended meaning. During comprehensibility test, all respondents should input their answers to the question: "What do you think this sign means?" If the respondent cannot determine the meaning of the sign, the input is "I don't know". In order to avoid the impact of the sequence on the test results, all safety signs are randomly set. Respondents cannot skip a page to answer, and must complete the test page by page.

2.5 Design Feature Test

After comprehensibility test, a design feature test was conduct to test three key design features of those safety signs. Key design features selected include familiarity, abstract, and relevance.

2.6 Procedure

The whole test is divided into two phases. In the first phase, the safety sign comprehensibility test is carried out. The main tester distributed the test materials of 6 safety signs to 100 respondents through the main control computer. The test time of each safety signs is up to 3 min. After the test page is shown, the respondents should input their answer as soon. After completing the test, all respondents rest for 10 min.

Then, the design feature test is carried out. Before the test began, the main tester introduced the meaning and evaluation criteria of the three design features to respondents, and the evaluation standard is familiarity (never – scored 0, often–scored 10), abstract (very abstract – scored 0, very similar—scored 10), relevance (meaning not related to design – scored 0, meaning strongly related to design – scored 10). Next, the main tester circulated the testing materials to all the respondents. On each testing page, there were one safety sign and its correct meanings. After seeing the sign, the respondents scored the design features of each safety sign. The test time for each safety sign was up to 3 min.

All respondents were required to complete the test independently and could not communicate.

3 Categorizing the Results

According to ISO 9186-1 [14], each comprehensibility answer was input and categorized into four categories (1, 2a, 2b, or 3) shown in Table 1.

Table 1. Categorization of answers

Category	Meaning
1	Correct
2	Wrong
3	Wrong and opposite
4	"I don't know"

4 Test Results Analysis

The data of each safety sign's comprehensibility, design features and their relationships was analyzed (shown in Table 2, 3, 4).

Table 2. Test results for 6 safety signs comprehensibility test

	Break to obtain access	General alarm	Lifebuoy	Wear a mask	Use handrail	Secure gas cylinders
1	86	62	59	92	98	38
2	12	29	20	8	2	25
3	0	0	0	0	0	0
4	2	9	21	0	0	37
Score	86%	**62%**	**59%**	92%	98%	**38%**

Table 3. Test results for 6 safety signs design feature test

	Break to obtain access	General alarm	Lifebuoy	Wear a mask	Use handrail	Secure gas cylinders
1	8	5	6	9	10	3
2	9	**3**	**4**	10	10	**2**
3	7	**2**	**5**	9	9	**4**
Score	80%	33.3%	50%	93.3%	96.7%	30%

5 Discussion

5.1 Comprehensibility Test

66% is the comprehensibility test acceptable score determined by ISO/TC 145. 3 safety signs scored above 66% were: "Break to obtain access" "Wear a mask" "Use handrail". 3 symbols with low comprehensibility were: "General alarm" "Lifebuoy" "Secure gas cylinders".

The average score was 72.5%. There were 3 safety signs scored above average level.

The average score of safe condition signs was 69%. And the average score of mandatory action signs was 76%.

Safety signs with a score of over 66% can be used without auxiliary text, but for safety reasons, it is recommended to have auxiliary text. Safety signs with a score of below 66% must be accompanied by auxiliary text to ensure the safety of people's lives and property.

5.2 Design Feature Test

The average score of the design features test was 63.9%, i.e., familiarity 68.3%, abstract 63.3%, and relevance 60%.

"Use handrail" got the highest score 96.7%, while "Secure gas cylinders" scored lowest 30%.

The safety sign got the highest familiarity score (full score 10) was "Use handrail". The safety signs got the highest abstract score (full score 10) were "Wear a mask" and "Use handrail". And those two safety signs also got the highest relevance score 9.

5.3 Improve Safety Sign Design

Familiar with Danger. To comprehend a safety sign, it is essential to improve people's familiarity with the danger. From the test, it was found safety signs with comprehensibility scores below 66% also scored lowest at familiarity, i.e., "General alarm" (comprehensibility score 62%, familiarity score 5), "Lifebuoy" (comprehensibility score 59%, familiarity score 6) "Secure gas cylinders" (comprehensibility score 38%, familiarity score 3). If most people are not familiar with the danger and the environment in which the danger exists, then early education and training should be adopted to improve familiarity. For example, in areas far from the danger, safety signs and auxiliary text can be used to repeat the dangers that may exist, and let people be familiar with the safety signs and related dangers.

Avoid Abstract Design. From the test, it was obvious that safety signs with comprehensibility scores below 66% also scored lowest at abstract, i.e., "General alarm" (comprehensibility score 62%, abstract score 3), "Lifebuoy" (comprehensibility score 59%, abstract score 4) "Secure gas cylinders" (comprehensibility score 38%, abstract score 2). When designing a safety sign, concrete images should be chosen to represent the actual object.

Relevance with Function. To design a standardized safety sign, it is essential to identify main danger. There are many potential hazards in daily life and work environment, and the design of safety signs needs to be in line with actual needs. Some hazards do exist, but due to insufficient actual demand, the comprehensibility test of safety signs will not have high score. The meaning conveyed by the safety sign should have close relevance with function. For example, high pressure gas cylinders should be kept away from open flames and electrical equipment when in use. While improper fixing is not a major danger when using high pressure gas cylinders.

6 Conclusion

From this study, it is suggested that when designing a new safety sign, designers should first identify the actual main danger, choose concrete graphical elements to design, and take care of improving familiarity when in use.

The test results showed that when developing a new safety sign, designers should pay more attention to the three design features which impact the comprehensibility greatly.

The comprehensibility of safety signs is also influenced by factors such as people's cultural background and education level. Therefore, in the future study of safety signs, it is possible to consider the factors of different respondents' cultural backgrounds and education level.

Acknowledgments. This research was supported by National Key R&D Program of China (NQI) (2016YFF0201700, 2016YFF0202806).

References

1. ISO 3864-1 Graphical symbols—Safety colours and safety signs—Part 1: Design principles for safety signs and safety markings. International Organization for Standardization (ISO) (2011)
2. Albert, A., Hallowell, M., Kleiner, B., Chen, A., Golparvar-Fard, M.: Enhancing construction hazard recognition with high-fidelity augmented virtuality. J. Constr. Eng. Manage. **140**, 04014024 (2014). https://doi.org/10.1061/(ASCE)CO.1943-7862.0000860
3. Dianye, Z., Rao, Z.: The passenger ship of inland waterway is choosing the simulation study for the emergency evacuation route. Comput. Simul. **01**, 151–154 (2018)
4. Rousek, J.B., Hallbeck, M.S.: Improving and analyzing signage within a healthcare setting. Appl. Ergon. **42**(6), 771–784 (2011)
5. GB/T 2893 Graphical symbols - Safety colours and safety signs. Standardization Administration of the People's Republic of China (SAC) (2013)
6. GB 2894 Safety signs and guideline for the use. Standardization Administration of the People's Republic of China (SAC) (2008)
7. GB 13495 Fire safety signs. Standardization Administration of the People's Republic of China (SAC) (2015)
8. GB 16557 Safety signs for life-saving on sea-going ship. Standardization Administration of the People's Republic of China (SAC) (2010)
9. ISO 7010 Graphical symbols—Safety colours and safety signs — Registered safety signs. International Organization for Standardization (ISO) (2011)
10. Yuan, J.: Empirical Study on the Factors Affecting the Effectiveness of Safety Signs. Zhejiang University, School of Management (2009)
11. Hu, Y., et al.: Evaluating effectiveness of safety signs on building site. China Saf. Sci. J. **22** (08), 37–42 (2012)
12. Chan, A.H.S., Ng, A.W.Y.: Investigation of guessability of industrial safety signs: Effects of prospective-user factors and cognitive sign features. Int. J. Ind. Ergon. **40**(6), 689–697 (2010)
13. Easterby, R.S., Hakiel, S.R.: Field testing of consumer safety signs: the comprehension of pictorially presented messages. Appl. Ergon. **12**(3), 143–152 (1981)
14. ISO 9186-1 Graphical symbols – Test methods – Part 1: Method for testing comprehensibility. International Organization for Standardization (ISO) (2014)

Research on Design Communication Mode Based on White Bi Theory

Xinying Wu(✉), Minggang Yang, and Xinxin Zhang

School of Art, Design and Media, East China University of Science and Technology, M. Box 286 NO. 130, Meilong Road, Xuhui District, Shanghai 200237, China
XinyingWu@qq.com, MinggangYang@qq.com, XinxinZhang@qq.com

Abstract. Chinese traditional aesthetic philosophy pursues "natural beauty", "essential beauty", "simple beauty" and "simple beauty". The aesthetic system in Chinese culture often ends up with a simple beauty. It can be seen from the Chinese poetry culture evaluation system that rhetorical beautiful poetry in Chinese history has never been the highest standard. As a representative work of Chinese poetry, the Book of Songs conquers the peak of poetry art. This is from the appreciation of beauty to the transcendence of beauty, but also from the process of understanding beauty to knowing beauty. Both the "category" and the "process" are expounding an ancient Chinese philosophical thought: the theory of "White Bi". The "White Bi" comes from the book of Changes, meaning no decoration. In the aesthetic system of ancient Chinese design culture, White Bi was regarded as the highest realm. In the end, it is still a natural concept of pursuing simplicity, turning the natural things and the universe and human factors into a state of harmony and unity.It has been more than 3,000 years since the birth of the theory of White Bi, representing the most traditional philosophical in China. Decoration is an essential element in design. The most important theory in the day is that the decoration will reach a pole and it will show a minimal state. Minimalist design is one of the manifestations of the theory of White Bi. The design of White Bi design has its unique way of communication. The design communication of White Bi is very important to human factors. This paper studies the design communication methods contained in the theory of White Bi. Relevant discussions were carried out in terms of human factors, culture, art, and communication channels. Under the premise of combining ancient Chinese philosophical thoughts, it puts forward a more effective reference method for the current design development.

Keywords: White Bi · Design communication · Human factors · Art design

1 Introduction

The divinatory symbols of Bi is from the sacred image of the Yijing, and it is the earliest ancient Chinese to explore the decoration. The ancients used the evolution of the image to illustrate the changes in the decoration in the design. The process of progression from the bottom to the top of the Book of Changes is a process of

A. G. Ho (Ed.): AHFE 2019, AISC 974, pp. 124–129, 2020.
https://doi.org/10.1007/978-3-030-20500-3_13

weakening the decoration [1]. It is "hexagram gen", and the bottom is (Fig. 1). In the explanation of the elephant, it represents the beautiful picture of the flames under the mountain. The truth contained in divinatory symbols of Bi reflects the definition and pursuit of beauty in Chinese traditional culture.

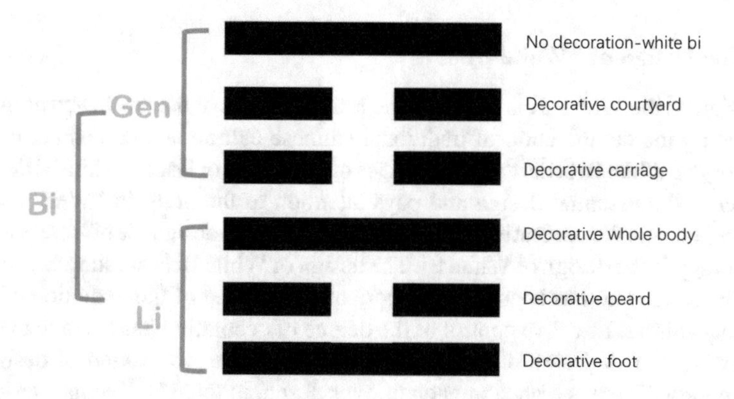

Fig. 1. The divinatory symbols of Bi

As the top layer of cockroaches, White Bi is a definition of decoration, and it is an emphasis on essence. The elaboration of the relationship between "Wen" and "Zhi" contained in divinatory symbols of Bi constitutes an important element of design thinking in traditional Chinese aesthetics. In the week of Zhouyi, the meaning of decoration is that the white enamel has no meaning in decoration in Zhouyi. It is also an image, but it has different meanings. This layer of evolution evolves from decoration to no decoration, which contains the beauty of returning to nature and returning to nature. The concept of White Bi aesthetics exists in the rich form of Chinese aesthetics and becomes an important part of Chinese culture. The theory of White Bi is rooted in the aesthetic system of the White Bi, and it is an important theoretical idea of the design of the White Bi. The theory of White Bi is embodied in the design category of simple design, green design and sustainable design. The simple design concept has become the ontology that constitutes the theory of White Bi. The theory of White Bi is also very important for the spread of contemporary design.

2 The Mode of Communication of White Bi Design

From a broad perspective of communication, design, like the speech act of people in daily life, belongs to the communication behavior of human society [2]. The intersection and integration of design and communication has become a new development in contemporary design development. The exploration of design communication methods can enrich the development of design culture and find ways and value rules that are more suitable for design communication. In the contemporary development of science and technology and the development of cultural diversification, designers should not only stay in the design route, design technology, design value, etc., but also further strengthen the human factor related design from the communication mode and psychological level. Let life and design become an organic whole. From a certain perspective, design communication is a type of cultural activity. Therefore, based on the theory of White Bi, the mode of its dissemination is carefully and smoothly, and a regular theory is summarized, which gives a meaningful enlightenment to contemporary design.

2.1 The Design of "White Bi"

The design of the White Bi is based on the artistic design of the aesthetics of the Bai bi, representing the combination of traditional Chinese culture and contemporary design. The theory of White Bi is the theoretical basis of the design of Bai bi. White Bi design pays attention to the essential design and pays attention to the unity of "Wen" and "Zhi". "Wen" represents the decorative ornament in the White Bi design. "Zhi" refers to essence and ontology in the design of White Bi. The design of White Bi is the simplification of the design language and has become an important design idea of the evolution of Chinese traditional culture. Effective control of the degree of decoration has become the primary rule of design. The design of the d White Bi can evolve into a kind of design in the contemporary. There is a phenomenon of over-design in today's designs. Designers are prone to over-engineering and creating barriers to design resources and user interaction. The White Bi design is the best way to solve this problem.

2.2 Mode of Transmission

The design communication method under the theory of c White Bi is very complicated, and it is summarized into the following main types. From the time division, it can be divided into three categories: art culture communication, regional space communication, and human factor communication. A special communication relationship is established between the designer and the consumer (Fig. 2). The theory of White Bi can spread to the present in the millennium, with a unique way of transmission. At the same time, it also proves the feasibility of the design of the White Bi. The design of White Bi is not only to effectively improve and innovate products, but also to transform design into effective information and into the communication process. Consumers are both an experience designer and an important component of the dissemination of human factors. Similarly, consumers are highly engaged in art culture and regional division.

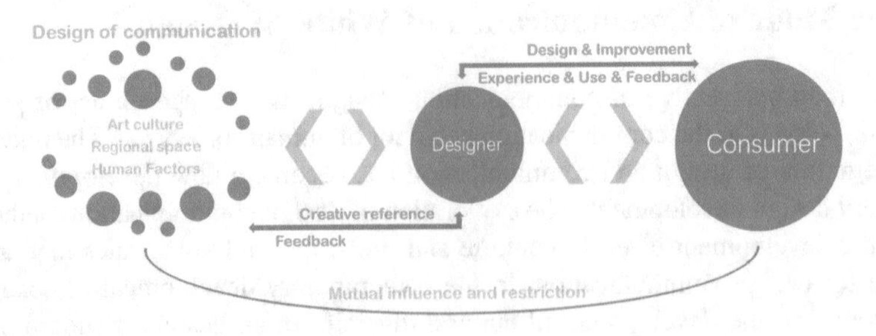

Fig. 2. Designer and consumer communication model

Art culture communication: Art culture is the superstructure of human cultural thought. In the spread of design, art and culture communication dominates. Design involves the fusion of art and culture, so the difference in culture, the difference in art genre will cause changes in the way design is transmitted. With the continuous advancement of society, the elements of artistic and cultural communication have also changed. Traditional elements include: artist, creative content, mode of communication, and popular aesthetics. Now these components can be transformed into each other and freely combined. The spread of the theory of White Bi has also changed with the change of art culture. From the history of Chinese art culture, it can be concluded that during the period of rich art and culture, people's aesthetics will gradually change from tedious to simple. Aesthetic taste will be greatly improved. In today's rich arts and culture, we advocate simple design, green design, and moderate design, in line with the ultimate concept of white design. The theory of White Bi has become a new direction to guide future design, and the design of white enamel will become the mainstream communication force of design.

Regional spatial communication: After human beings enter civilized society, natural geographical division leads to the formation of regions [3]. Regional culture began to take shape and regional design was born. Regional communication has also become one of the main types of design communication. The design communication of the White Bi is spread in a special regional space. In the same regional space, the cultures will be relatively the same, and the channels of communication will be relatively similar. Then the design communication method will form a similar pattern. However, today in the rapid exchange of information on globalization. The regional space of the design has been gradually reduced, and the spread between the regions has also become closer. Simplifying the design language helps to spread the exchange of information. Although the spread of design continues to change, the theory of White Bi design has always been unique. The theory of White Bi is rooted in the ancient Chinese philosophical system and aesthetic system. The uniqueness and unrepeatable characteristics of the White Bi have become the particularity and feasibility of the theory of Bai Bi in the regional space.

Human factor communication: The human factor determines the spread of design. The way human factors are transmitted is multi-faceted and multi-dimensional, and a large part of them are in people's aesthetics and feelings of use. Complex designs can bring a variety of functions and usage feelings, but they may not bring the best use experience and function utilization. The design of the White Bi promotes green design, minimal design, and proper design. This design is in line with the development of the times. The design of White Bi under the support of the concept of White Bi has a wide spread of power in modern society. In the context of the new media era, the way human factors are transmitted is reflected in the design content and consumption value. Design and art have become the highest form of expression of people's ideology. The higher the level of social civilization, the higher the level of people's ideology will be. The aesthetics of design pay more attention to environmental protection and nature. These are in line with the theoretical advocacy of the design of the day.

3 White Bi Theory Design Communication Model

In the mode of communication, Maletzk's mode of communication has been accepted by society. In this mode of communication, the four major components of the communication structure have not changed, and the complex relationship of each element has been detailed. According to the design communication mode under the theory of White Bi, it can be concluded that the design pays attention to the objective factors such as culture, region and value in the creation, and also pays attention to the human factors based on the Maletzke communication model, according to the characteristics of the theory and design of the White Bi, for new model making (Fig. 3). This model can provide a new reference for designers. Elements in the communication model can also be a design element for designer innovation and improvement.

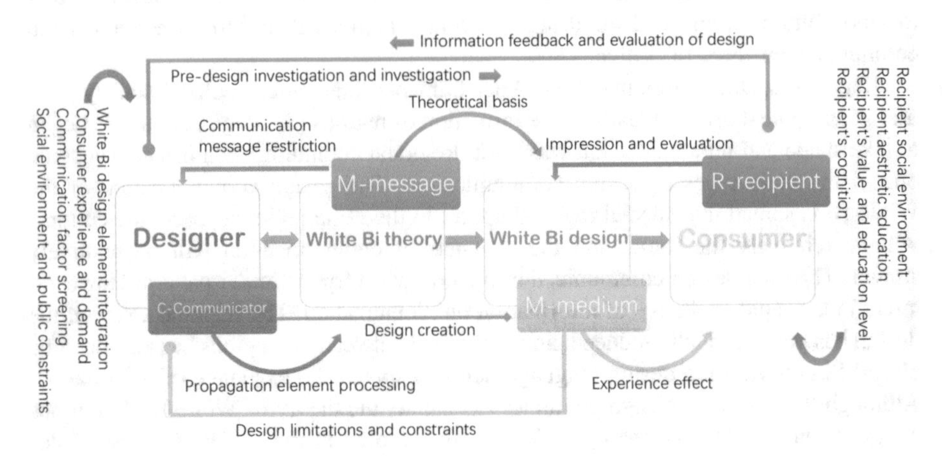

Fig. 3. White Bi theory design communication model

4 Conclusion

Combined with the modern design method, through the redefinition of the theory of White Bi, let the design of the White Bi return to the contemporary design with new elements. The sustainable design, green design and simple design advocated by White Bi theory meet the design needs of the modern era. It can be concluded from the conclusion that the White Bi design propagation mode under the theory of Bai Bi has certain regularity. Combined with Maletzke's mode of communication, combined with the propagation mode of White Bi design, a new propagation model is obtained. It helps designers to better design innovatively based on communication factors. Provide a more detailed guidance program for the development of the theory of White Bi. The concept of White Bi has spread to the present, and it still has the advanced design theory after thousands of years. Studying the new communication model while providing a reference for designers, it can make the design of the day with innovative significance and more suitable for the requirements of contemporary design elements. Ability to propose new standards and considerations for design.

References

1. Wu, X.: Analysis of the architectural art of "Bai Bi beauty" context.D. Henan University, pp. 7–8 (2015)
2. Xu, Z.: Analysis on design's context connotation from the communication perspective. J. Shenyang Aerosp. Univ. 172–174 (2017)
3. Li, L.: Value of design culture and its cultural dissemination. J. Art Design, 8–9 (2005)

3D Scanning and Visual Dimension – Technological and Creative Evolution

Man Lai-man Tin(✉)

School of Arts and Social Sciences, The Open University of Hong Kong,
Ho Man Tin, Kowloon, Hong Kong
lmtin@ouhk.edu.hk

Abstract. Humans have been trying to imitate and represent the world visually since ancient time. With the development of technology and computational visualization, the transformation of the imaging techniques from conventional media to 3D scanning as a new representation method with digital data generated has been widely applied especially in the creative aspect in recent years.

3D scanning technology provides us a new insight to understand the formation and essence of photographic imaging, in which the 3-dimensional image formation provides us with more layers to understand the sensory and knowledge about time and space, cognitive of being, and the technological and creative evolution in art.

This paper attempts to discuss the essential transformation in dimensional imagery with human factors and its aesthetic experiences through digital art means and philosophical approach.

Keywords: 3D scanning · Dimensions · 2D image · 3D model · Being · Time and space

1 Introduction

In the 21st Century, we find ourselves living in an increasingly globalized and multimedia world where images have been widely applied especially in the creative aspect and pervade the communications of our everyday lives. Images have been a dominant communication form since ancient time, they are being remixed, reformed and reformulated across a range of media, formats and cultural contexts across the history of evolution. One obvious development in imaging technique in recent years is the transformation from conventional 2D visualization to 3D imaging methods, in which the works or products under technological and imaging transformation through 3D scanning as a new imaging and representation method have been applied and presented in both the commercial industry and creative arts areas. However - we are not able to understand and interpret the changes and interrelationship among images, visualization and technological transformation based on previously mentioned lateral shift. Such phenomenon is actually the result of a long accumulation of aesthetics and optical technology evolution. By examining the contemporary image production and its presentation possibilities, it is hoped that a relatively complete "image landscape" can be

A. G. Ho (Ed.): AHFE 2019, AISC 974, pp. 130–137, 2020.
https://doi.org/10.1007/978-3-030-20500-3_14

unveiled. This paper will discuss the essential transformation in dimensional imagery with human factors and the aesthetic experiences through different examples of works.

2 Imagery History

Cave paintings in Africa, Europe, Australia, and Asia were made around 40,000 BCE. This prehistoric artwork present the first attempt of human to represent what they see through 2D visualization. The figurative art paintings are mainly the imitation of animals. The forms, shapes, and details of the animal paintings reveal a long observation and understanding of the characteristics of the 3D world. To understand the imitation ability of human, we need to first understand the optical mechanism - our eyes are separated on the face which allows the retina produces 2D images with some deviations. Such imagery differences create optical illusions to help us understand the depth of the 3D world. Human thus can interpret the 3 dimensions of things and space through temporal observation and sensory experience. Such ability in a way can be regarded as a kind of biological 3D scanner which can produce stereoscopic images and relevant perception about 3 dimensions.

In 1908 when we first found the figurine, Venus of Willendorf which has been created about 30,000 BCE, it provides us evidence of evolution about 3D visualization and replication of 3D world from painting to plastic arts, human's cognitive level about 3D visualization has been escalated.

Cave paintings and Venus of Willendorf are not just significant art pieces in history, but also a key for us to understand the two fundamental human factors to replicate 3D world in art - 2D image and 3D modeling. Since then human started to develop techniques, technologies, imaging apparatus and possibilities in making 2D images and 3D models, different approaches such as painting, sculpture, ceramics, photography and computer graphics, etc. have been created. Among these methods, 3D scanning is an emerging imaging technology and method. 3D scanning was first created in the 1960s. Laser and shadowing techniques were introduced into the scanning technology in 1985. Since then 3D scanning devices take different key technologies with the rapid development of image processing systems. Around 2005, digital 3D replication technologies developed by PrimeSense and Microsoft provide incredibly fast and flexible reality capture system to integrate 2D images and 3D models, allowing motion capture and interaction between devices and users. After more than a decade development, digital 3D replication technologies have been widely used in commercial, industrial and medical applications, the technologies have also been extensively used in artistic and cultural areas in recent years.

3 3D Scanning Technology

Besides a certain degree of resolution and details can be generated, 3D scanning can process higher flexibility for the creative approach in terms of scanning manipulation, digital images and geometry data generation. Many 3D scanners are available as portable handheld devices, digital image and geometry data replication become more flexible and easy to handle. The digital data are available to extend to other form and media such as 3D printing and mobile application. This implies a lower cost in visualization in evolution [1] and thus there is a significant growth in use and application of 3D scanning in last few years.

The data captured through 3D scanning is a replication of 3-dimensional space, the digital images and geometry data are described as point cloud in the form of x, y, and z coordinates. This Euclidean geometric presentation allows human to understand and interpret the digital 3D representations of reality. In general, there are three types of 3D scanning, namely Photogrammetry, LiDAR, and Infrared.

3.1 Photogrammetry

Photogrammetry in 3D scanning refers to the technology of making the measurement from photographs. A precise setting with cameras (normally using SLR digital cameras) from different angles is needed for taking photos. The multiple photos taken instantaneously will be calculated by software to generate the geometric data and 3D model. The image resolution, therefore, depends on the quality of the cameras and the preciseness of the setting. The drawback of this type of scanning is comparatively high in cost due to equipment, restriction of location and low portability for scanning.

3.2 LiDAR

This type of scanning is based on the light measurement of reflected pulses between the device and the scanned object by laser and sensor. High resolution, details, and accuracy of 3D data and images can be generated. This scanning method normally requires more expensive devices and specialized knowledge in operation.

3.3 Infrared/Structured Light Scanning

The cost of this technology is relatively low, in comparison with Photogrammetry and LiDAR. Like PrimeSense Carmine and Microsoft Kinect, the 3D scanner generally consists of cameras, an infrared projector and a CMOS image sensor for the collection of surface data and distance information, and extracts the 3D data in the scanning field within its own system. Since infrared scanning technology is available to implement through a portable handheld device, the 3D scanning process can take place in any location, which provides high flexibility and mobility for set up and scanning. However - multiple layers of scan data of the same part of scan object or area will be created due to multiple scanning processes. This will reduce the accuracy of data and image, which creates a drawback of this type of scanning.

4 Reality Replication and Representation

3D scanning is one of the best technologies for reality documentation in view of its ability to digitize things with high resolution and capture precise surface data and distance information in a relatively fast and easy way. In 2014, former President of the

United State Barack Obama was being 3D scanned by using a set of custom-built LED lights and two handheld structured light scanners in the White House. A high resolution of digital 3D Barack Obama portrait was captured and transformed into a 3D printed sculpture. This was widely reported in the media and regarded as the next generation of 3D portraiture.

One year before Obama portrait sculpture was made, Marina Abramović, the world's most famous performance artist, first presented her work "Five Stages of Maya Dance", a series of portrait sculptures combine performance, rendering technology, light and tonal. Through 3D scanning and computational rendering, her performance with strong facial expressions was scanned. A series of digital 3D tonal portraits (Fig. 1) that can rotate the viewing angle (mainly 180°) were milled in five alabaster blocks (Fig. 2). Both works present the possibility of transforming 3D scan into solid material, and the extension of digital data into a materialized art form. To an extent, "Five Stages of Maya Dance" explores the possibility of reinterpreting performance - the alabaster blocks as a physical presence of Marina Abramović allow the audience to observe the changes of light and tone created by 3D printing, which transform the facial expressions into Eadweard Muybridge's style-like stop-motion photographs. This work attempts to blur the boundaries between photography and sculpture, stop-motion and performance.

Another artwork of Marina Abramović "Rising" (2017) further explores the possibility of communicating and interacting between performer and audience in virtual reality. The high resolution and facial detail captured through handheld and underwater 3D scanner make it possible to build a virtual portrait of Marina Abramović in a fast and a highly convincing way. The 3D scan data were manipulated and recycled virtually and presented in an immersive environment, which allow her imagination between reality and fantasy to expand. The work also explores the digital humanity issue as mentioned by Marina Abramović that she is interested in how the digital 3D Abramović and her VR performance will affect the audiences' view on climate change and its impact on the real world. Marina Abramović's work has a dimensional and representational significance which can shed some light on 3D scanning in art.

Fig. 1. 3D scanned Marina Abramović portrait with tonal value and "landscape" generated.

Fig. 2. One of the finished alabaster pieces of the work titled Five Stages of Maya Dance. The series of works were presented at Masterpiece London 2018.

Frederik Heyman, a Belgium artist and photographer, shows his final fantasy of memories through Photogrammetry. Based on his illustration and photography background, he extends his artistic idea to the digital 3D world. Photogrammetry as one of the major 3D scanning methods, it can generate precise geometric data and 3D model with high-resolution texture and surface data. This allows Frederik Heyman to create highly realistic imagery and surreal stage composition - a sophisticated digital 3D world composed of illustration, installation, photography and video. Each element and movement in the work has been carefully designed and planned to break through the 2D narrative with a strong sense of mise en scène. The scanning process and digital 3D model movement in post-production provide the observer with a temporal dimension to experience the 3-dimension and expand the narrative possibility. 3D scanning can integrate the subjective sensory experience and objective presentation of scanned things. This challenges the traditional imagery methods through photography and computer 3D modeling.

Humans' imagination of reality and fantasy have been continuing since the prehistoric art 40,000 BCE ago which can now be expanded to a new dimension through 3D scanning. In the Instagram account titled "brain___________________heart", there are many stills and videos by using 3D scanning technology. The artistic approach of these works is quite different from Marina Abramović and Frederik Heyman as scan noises are remained in the work deliberately together with the main objects or scenes (Fig. 3). Some of the scan objects are even presented in an imperfect or fragmented form. This shows another characteristic of the 3D scanner in creative approach - creating "error" or "misinterpretation" data in scanning. The 3D scanner and relevant software do not filter the noises and decide what should be reminded in multiple scans automatically. By ignoring the usual practice of scanning post-production such as "remove parts" or "fill holes" to a certain extent, the works remain their most organic and original state of scan with parts or forms that are not intended to scan originally, which expand the concept of aesthetics experience - from replication of real world to beyond the visionary world. This reminds me of the artwork by Henry Moore, an English artist who is best known for his semi-abstract sculpture that allows audiences to see both positive and negative space, the fragmented data and noises presented in "brain____________________heart" can themselves have as much meaning as the main scan objects. The concealed elements are revealed which engage the audiences to interpret the work in a role of control, the meaning is destabilized and expanded disregard of original narrative structure. The works have writerly texts [2] characteristic emphasised by Roland Barthes, a French literary theorist, philosopher, critic, and semiotician.

The scanning process is also significant for the understanding of fundamental concept about "Being". From the result and the visualization style of the scan data appears in the works of "brain____________________heart", it is very likely that the works were created by infrared/structured light scanner. By capturing full-color model through 3D scanning and represent the model in digital means, the scanning process, especially by freehand scanning, examines the shifting of the mode of "Being ready-to-hand" to "Being presence-at-hand". The object is being examined carefully and intentionally during the scanning process. This is an intrinsic shift ready-presence philosophical cognitive pattern, based on the theory of the German philosopher Martin Heidegger [3].

Fig. 3. Screenshots of video show the fragmented data and noises captured by 3D scanning.

In 2015, Antoine Delach, a French CG artist, released his stereoscopic short film "Ghost Cell" (Fig. 4) which was made by using laser and photogrammetry 3D scanning technologies. The city of Paris was transformed into one of the fundamental visual elements - line (wire) in a dream-like atmosphere. This work is significant for its escalation of "error" scanning possibility. Things appear in the film are composed of sophisticated and intertwining wires. The misinterpreted data interlace with moving images collected by 3D scanning creates not only a new perspective to understand the 3D world, but also reshape the temporal and spatial dimensions and the awareness of the fundamental concept about time and space, the digital data were collected and represented as a new narrative of time and space.

Fig. 4. Ghost Cell, 2016. By using laser and photogrammetry 3D scanning technologies, the city is transformed into sophisticated and intertwining wires with fragmented information and visual effect.

5 Conclusion

The fundamental human factors to replicate the 3D world in art - 2D image and 3D modeling have been addressed by 3D scanning technology. 3D scanning not only as a mean to replicate the reality with digital data generated, but also a new insight for us to understand the formation and essence of images and dimensions. Perhaps it is time to review Plato's question about the "shadows" and the true form of reality raised in his Allegory of the Cave [4] through 3D scanning. 3D scanning in art can shed some light on the temporal and spatial dimensions, which provide a new perspective of observation, cognitive of being and narrative possibility. The 3D scanning art like "Ghost Cell" are echoing with the cave paintings and Venus of Willendorf 30,000–40,000 BCE ago, this is a new chapter of technological and creative evolution of human history.

References

1. Dawkins, R.: The Selfish Gene: 30th Anniversary. OUP Oxford, Oxford (2006)
2. Barthes, R., Miller, R., Howard, R.: S/Z. Blackwell, Malden (2002)
3. Heidegger, M., Schmidt, D.J., Stambaugh, J.: Being and Time: A Revised Edition of the Stambaugh Translation. State University of New York Press, Albany (2010)
4. Plato, J.B.: The Allegory of the Cave. Enhanced Media Publishing, Los Angeles (2017)

5. Fig. 1. 3D Render of Five Stages of Maya Dance (2018). [video] https://vimeo.com/257534261. Accessed 9 Mar 2019
6. Fig. 2. Five Stages of Maya Dance, 4 of 5 (2013). [image] http://www.factum-arte.com/pag/115/Five-Stages-of-Maya-Dance. Accessed 9 Mar 2019
7. Fig. 3. Saint ursula and her handmaidens (2019). [video] https://www.instagram.com/p/BpK7fDmFOHd/. Accessed 9 Mar 2019
8. Fig. 4. Ghost Cell (2016). [video] https://vimeo.com/139651679. Accessed 9 Mar 2019

Communication in Design

Kids at Preschool. Designing Products and Wayfinding Systems to Enhance Kids' Skills, Facilitating Wellbeing Through Communication

Laura Giraldi(✉), Elisabetta Benelli, Marta Maini, and Francesca Morelli

University of Florence, Florence, Italy
{laura.giraldi,elisabetta.benelli}@unifi.it,
marta.maini@hotmail.it, francesca_morelli_@libero.it

Abstract. In our society kids spend a significant part of their life outside home up to eight hours a day there. In this environment, to feel safe kids "need" both rules to respect and freedom to live experiences. Still today pre-school are designed mainly according only to the ergonomics rules without considering emotional aspects. The main objective of this research is to identify a set of good practices to design pre-school products, furniture, interior settings and wayfinding systems able to enhance quality experiences on education and growing to integrate learning and teaching methods in pleasant experiences, improving awareness among the users. This paper aims to underline the importance of the communicative aspects of design for a new education approach based on "learning by doing".

Keywords: Children · Pre-school design · Kindergarten · Environment · Learning by doing · Communication · Emotion · Education

1 Introduction

1.1 Context of Reference

Nowadays, the educational tendencies in use refer, mostly, to educational models introduced in the past, based on principles such as relationship, experience, autonomy and security. Among the most widespread we underline the model of Maria Montessori based on independence and autonomy.

According to her theories, the educator's role is not to teach or judge children, but to help them, to be responsible and autonomous guiding them in a place that it is not competitive but collective. She used to say: "let's help them to do it alone" [1].

Really, the educational model does not only concern teaching, but it also refers to the surrounding space: "The environment must be welcoming [...] without superfluous things [...]. The furnishings must be child-friendly and all the spaces pleasant and welcoming [...]. The environment must be flexible for different work's activities and at the same time children have to adapt it by preparing it for their future work" [1].

A. G. Ho (Ed.): AHFE 2019, AISC 974, pp. 141–150, 2020.
https://doi.org/10.1007/978-3-030-20500-3_15

The Loris Malaguzzi model, also known as *Reggio Emilia Approach*, is another very rife educational model. In addition to the previous educational model, he takes care of their moods, preferring the relationships and the interactions among children and their natural behaviors during workshop and multimedia activities.

He affirmed that "The child is made of a hundred languages, a hundred hands, a hundred thoughts, a hundred ways of thinking, of playing and speaking, a hundred ways of listening, [...] a hundred worlds to discover, a hundred worlds to invent, a hundred worlds to dream" [2].

The way to express children‘s emotions is through the activities carried out in the ateliers, spaces dedicated to artistic activities, where the expressive and poetic languages generate the necessary knowledge for the development of the child.

According to these models, the daytime in the kindergarten is divided into routines, dedicated to different tasks. The corresponding spaces must be designed in a way to guarantee all users' needs and facilitating the correct carrying out of all the activities.

Scanning the day in routines means teaching the child the behavioral rules that will accompany him throughout their present and future life.

1.2 Interior Settings and Functions at Preschool

Still today preschool is generally set in a traditional way as it was twenty or thirty years ago. Spaces and products are designed according to ergonomics rules, including measurements, furnishing, materials and surface finishes for safety [3].

Instead, it is also necessary to consider children taking into consideration their emotions. Despite the growing interest for this issue, which encompasses different disciplines (i.e. design, pedagogy, psychology) and despite the several international prizes on designing new buildings for pre-school, nowadays it is very difficult to find interiors specifically designed for them.

Generally the elements of pre-school interiors (furnishing, way-finding systems, graphics) are designed according to security requirements and anthropometric measures.

Regarding the distribution of spaces and functions for a long time the preschool was conceived as a unique place, developed in classrooms, while the other spaces had not importance because they were considered as subordinate to the centrality of the classroom. Today pedagogical studies on educational environment have proved the positive impact of spaces, products and furniture on kids' behaviour.

Different rooms and activities areas are furnished and equipped to foster discovery, exploration, social relation and interaction while promoting language, skills, creativity and the ability of problem solving.

Therefore, all the spaces should be organized to respond children's needs, they should be well characterized, differentiated and readable, and also safe and stimulating places to create a path of familiarization and knowledge that helps children to orientate spending pleasant time inside the kindergarten. The kindergarten is generally located in a building that welcomes children dispensing a variable range of services.

Preschool spaces are generally divided in the following main functional areas:

1. The entrance/welcome area
2. The classroom

3. The playground
4. The sleeping/nap room
5. The refectory
6. Daycare bathrooms
7. Connective areas (corridors, stairs, halls)
8. The courtyard

Furthermore, sometimes there are also gym areas and common spaces dedicated to events and end of year performances. According to children's age there are different kind of furniture that is adapted respect to the specific measures and sometimes to the skills of the children.

The research focuses on the welcome area and the classroom, a flexible space were children live and experiment structured and free activities.

Welcome Area. The welcome area is a space, generally located close to the entrance of the school, representing the meeting point of educators, parents and children. This area has two main functions: the practical one contains kids' personal belongings such as clothes, shoes, toys; the second one, more important, facilitates the separation of children from parents and from their familiar environment. Consequently this area hosts daily rituals that develop the autonomy of children and welcome moments of school-family communication.

Classroom. The classroom is the main environment of the kindergarten where structured, semi-structured and free activities take place, including spaces for playing and for laboratories. During the activities children learn how to relate to the other children, socialize and live their first experiences developing skills towards their autonomy. It is important they have every activity carried out without judgment, in order to stimulate personal initiative. During structured activities children usually sit in common tables of 4/6 places, which are often adjusted ad hoc for individual or group work.

Playground. This area is often included in the classroom area but sometime some kindergarten has also a dedicated space. The playing activity is an expression of child's creativity that learns and refines his skills through experience. During this activity the child is completely free to move, to express oneself and to relate with those who prefer, while the teacher has the task of observer and can only be involved upon the child's invitation.

The environment, the furnishing, the everyday objects, the graphics and all the settings elements became strategic to design a pleasant and educational environment for children and educator.

1.3 Children and Parents' Stress for Attending Preschool

The start attending pre-school often causes stress and worries in children and parents since it represents a determining step for the correct psychophysical child development. The hardest transition for children is the necessary shift from being at home to entering the classroom while parents are worried for the unpleasant emotion of their children in separating from them and their habits. Moreover they are also stressed to entrust their children to extraneous educators in an unknown environment. Educators can, however, help children avoid the emotional stress, they feel during this time, by adding some welcome activities to their routines. Stress is a concept scientifically defined and an evincible phenomenon [4].

In 1998, Evans and Mitchell McCoy described the way a built environment may influence not only the health of persons but also behavior and found which architectonic elements may lead relevant stress in the users. In pre-school environment, the stress did not allow children to feel at ease influencing their behavior and comfort. According to Ulrich, Devlin, Arneill and Del Nord, the elements of the environment that may represent possible stress vehicle for users are the following: the image itself of the spaces, sensorial feelings, difficulties, impossibility in control and manipulate the environment, orientation difficulties, physical discomfort.

2 The Aim of the Research

The aim of the present research is to propose an innovative approach to the pre-school design. The research proposes a series of good practices to design pre-school products, furniture, interior settings and wayfinding systems able to enhance kids' quality experiences on education and growing. This study proposes a new collaborative and innovative approach in designing for preschool spaces to integrate learning and teaching methods in pleasant experiences, improving awareness among the users helping, at the same time, educators during daily routines.

Moreover the purpose of this work is to underline the strategic role of design, and in particular its communicative aspects able to transfer material qualities and intangible values suggesting behaviors, relations and interactions among persons and between persons and products. In order to improve the wellness of kids and make them feel at ease in pre-school spaces, it is necessary to render those ones more friendly and familiar.

The study aims, indeed, to guide designers to design interiors, furnishings, products and way-finding systems able to orient, inform, interact, entertain include all kind of children using a universal language coming from their collective imaginary. To improve this kind of design culture, the present study recommends a methodological referring system to be easy applied in different indoor places dedicated to children. The suggested good practices are designed taking into account the peculiarities (skills, emotions, behaviors) of children at different ages in accordance to the theory of Maria Montessori [5].

The main goal of this work is to propose a referring methodological system to design products for kids in kindergarten environment able to communicate with them according to two different levels:

1. The first communication level is designed to orientate children guiding them along interiors and making them familiar. As a consequence children feel themselves at ease. This kind of communication is usually based on children behavior and corresponding emotions so that they recognize both pleasant situations and furniture and object to be used for a particular tasks.
2. The second communication level, stimulating children to interact with the objects and the environment and to relate to other children, allow them to live conscious

and educative experiences. These kind of communication is based on learning by doing through interactive and relational experiences children develop new skills and behaviors.

Summarizing the final research's aim is to find open rules and good practices for the design of kindergarten environment, flexible adjustable interactive, able to look familiar and to orient children stimulating pleasant and didactic experiences.

3 Methods of Research

3.1 Methodological and Multidisciplinary Approach

The research refers to the Human Factor approach in which children represent the key actors. To study children as main users, the design research used a multidisciplinary approach involving many actors from different disciplines as educators, psycho-logists, pedagogists, and even involving parents and children. The research starts from the studies of two Italian pedagogists Maria Montessori and Loris Malaguzzi, whose theories are the most followed all over the world. At the same time the present work uses the co-design method involving children in co-working to focalize on project priorities related to their unexpressed needs and emotions. Furthermore the work refers also to the education method based on "learning by doing", due to the practical approach centered on the material involvement of children in the environment and it works as a continuous training.

According to the Italian National Guide Line [6], pre-school educators should choose and propose children's activities to develop their specific skills. The present tools at their disposal today do not facilitate these tasks. Consequently, the design of kindergarten's environment (such as furniture, wayfinding systems, communication and products of daily use) plays a very important role in education, as the real responsible of the pre-school quality together with the educators. Starting from the consideration that children (3-6 years), according to their level of psycho-physical development, have their own skills and behaviors, it is necessary not only to study their needs before designing for them but also to outline a set of open rules for designer to take into account when designing for children.

By following the educational approaches we consider their indications about the environment of children life as a fundamental rules able to influence their development and growing [7]. Consequently, the design of children's pre-school environment (dealing with furniture, objects, graphic and interiors' element) needy to define a set of "basic elements" as, for instance, colors, graphics, shapes, helping designer to design according to children material and emotional needs.

Referring to children as main user it is important to use languages that they can easily understand and recognize, in designing for them. This is possible if we refer to kids' collective imaginary, necessary for capturing their interest and stimulating their skills. The research used also a co-design approach that has been experimented as described below during the development of the present work.

In particular, the research used the co-working activities for involving small users during design process.

3.2 Design Thinking and Co-working with Children

The focus of the co-designing process is not only working together in order to product output, but also to enhancing the design skills of the participants. It is very important for the designer to know the users and particularly their material and emotional needs, habits, behaviors and all the practices of their daily routine. The main users of this research are the children. In our society adults make all the decisions for them, do not very often thinking at their real needs. The child so becomes just a passive actor of the society [8]. Moreover, sometimes children are treated like little adult, not considering their physical and emotional needs, while it is important to provide them the best way to communicate and to encourage creativity making them feel at ease, also during the workshop. Infact the co-working activity, as a practice of *Design Thinking*, is a creative way to communicate with children understanding their "needs" and desires.

In recent years children are more and more included in different phases of the design processes in various ways. Allison Druin in 2002 said that a child could have four different roles during the design process: it can be a user, a tester, an informant or a design partner [9]. Smith and Dindler have introduced on Druin's model the role of the "protagonist" that empowers the child more [10], as theorized also by the Reggio Emilia approach to early childhood education. Fenne Van Doorn expanded the role of the child in a design process, by introducing the role of co-researcher [11]. In this case the child is both the researcher and the user. During the co-working activity with children some important factors should be ta-ken into consideration: the children age, the environment and the activity.

Given that abilities and skills of children change according to their different age [12], it's necessary take this factor into consideration since it's fundamental for the choice of the activity to be adopted during the co-working.

The environment in which the co-working is carried out is the second factor to take into consideration. For this reason, the choice of well-known places such as schools may be reasonable.

The third factor, which is strictly connected to the other ones, is the activity. As theorized by Druin, the use of low-tech methods is a good practice to encourage the child to communicate with them. Paper, clay, Lego, crayons can be considered as examples of low-tech methods. It is important do not frighten and do not judge the child but make him feel at ease. In this way, the child can feel free to express himself. The co-working activity should be ludic and, in the same time, should make the child responsible for the project.

For the purpose of this research, two different co-designing examples have been carried out, involving children of different ages and in similar environments. In both cases, were executed elementary activities characterized by the use of paper, crayons, marking pens, stickers and stationery.

4 Applied Method

The research studied the following aspects, by relating them with the specific scenarios of the Italian pre-school:

1. Study of the literature: perception psychology, children skills and behavior at the age of 3–6, way-finding systems, design for children, communication design, human factor design.
2. Direct observation of children at pre-school - as explained in the use case below.
3. Interviews to educators, and parents for detecting problems of welcoming, activities and orientation both for them and for children in pre-school.
4. Co-working activity with children aged 3 to 6 years.
5. Collection and comparison of results.

This practical activity of direct observation of children was based on the detection of their mood and needs when arriving at preschool, during daily routine, experimenting educational activities and playing.

The purpose of the co-working activity is also to know if and how children feel at ease and if they have difficult to orientate at preschool. Moreover, the experimentation aim to identify possible real solutions to designing furniture and products at preschool interpreting children suggestions and at the same time helping educators during their work with children.

The results have to find and/or verify a set of good practices to design preschool products and environment that are "children centered", using colors, shapes and familiar elements according to their own language and collective imaginary.

4.1 Use Case: Co-working Design

The experiment was set-up in Pontedera and Ponsacco, two small Italian towns close to Florence and involved eight pre-schools of mixed classes (aged 3–6) for a total amount of eight sections (one for each school) and 166 little students. The research has implemented three main actions that allowed to understanding better children behaviors:

1. Direct observation at welcome time and direct observation in classroom during different activities.
2. Interviews to educators.
3. Co-working activity with children.

These following actions were carried out within three months from September to December 2017.

Direct Observation. The observation started considering separately the entrance/ welcome area and the classroom. The activity has involved children aged from three to six years old. The observation showed how all the children were interacting with the environment and interior settings and how they were moving in the surrounding environment with and without their parents. The method's application points out, for the most, the difficulties of all the children to be oriented and feel comfortable in September during the beginning of the scholastic year. The observation also underlined that differently aged children common activity is the playing. Besides, the observation showed how children used to play with well-known and familiar elements, even if they were not designed specifically for playing.

Interviews to Educators. For the research purposes, questionnaire and interviews related to the functional areas mostly experienced by children were conducted with the educators. Fifty educators from eight kindergartens have been involved. The eight pages length interview covered the aspects related to shapes, materials, dimensions, dispositions and functions of the products and the furniture. Further, also tried to question possible improvements of welcome area, the classrooms and the playground. Educators reacted positively to this initiative, pointing out the problems related to the kindergarten and offering solutions for the design of the environment. They highlighted that the present furniture is often dated, not easily usable, even not able to stimulate and increase the physical and mental development of children. Moreover, the need of a better interaction between children and pre-school environment was particularly underlined.

Co-working with Children. The co-working activity with children was carried out in sessions of an hour each. During the co-working the children, followed by designers and educators, were asked to draw their ideal furniture for the school and that could "make them feel at ease", using the materials they preferred. The co-working activities have underlined useful aspects and the main characteristics regarding to shapes and colors to be used later in project. The work with children allowed the researchers to understand their thoughts and emotions related to the preschool environment, and their needs related to the furniture. The co-working activities underline that children were mainly attracted by colorful furniture, simple and rounded shapes able to stimulate their creativity and make them feel at ease. During the sessions children have highlighted the material use importance of products in pre-school environment, drawing, for instance, furniture made by soft and pleasant touch material. Besides, they have drawn equipped and technological furniture, able to transform themselves, to be multifunction and adaptable to the needs.

5 Results of the Research

The research states that all the elements inside a preschool environment as, furniture, objects, graphics, communicates material meaning and immaterial values, for this reason they have to be designed by specific characteristics to absolve explicit purposes. These characteristics have to come from children's imaginary, because they need to learn through friendly objects and playful elements, during ritual activities of their daily routine.

As a result, the research proposes a series of good practices applicable to all kind of kindergarten environment, both for new and for existing ones. The following points summarize the proposed actions useful to design this kind of places at "children size".

The research underlines the need to set specific goals to material elements at kids' disposal according to educators (and parents) requests and to children personality and feels. The starting point is to associate a series of values to each "functional" environment at preschool.

The results propose two main sphere of referring for kindergarten environment:

1. Reassuring environment
2. Educational environment

Each of the above design environments have to contain a series of specific elements necessary to obtain the prefixed goals.

Peculiarities of *reassuring environment*:

1. *Protective:* the child feels at ease in a safe environment through comforting and protective elements. Design elements' peculiarity: comfortable, cosy, pacify, realized by light colors, rounded shapes, small dimensions and soft touch materials.
2. *Recognizable:* the child is able to move in security and freedom orienting himself in spaces through ritual activities, familiar and customized elements. Design elements' peculiarity: ritual, familiar, from their imaginary, customized, realized by simple shapes, primary colors, easy to identify among others.

Peculiarities of *Educational environment*:

1. *Interactive:* the child is invited to interact with ludic aspects and with the elements' transformability. Design elements' peculiarity: visual, sound and light feedback. Use of simple and friendly shapes, soft materials, touching finishing, bright colors, intuitive graphic, multimedia elements.
2. *Relational:* the child is embroiled by functions, objects' shapes and their multi-functionality, especially if they work correctly with more than one user, stimulating and increasing collaboration, inclusion and relation-ship. Design elements' peculiarity: multifunctional shapes suggesting collaborative actions, simple info-graphics.
3. *Experiencing:* the child is involved by the environment, the furniture, the objects and all the material elements able to communicate information and inviting to the exploration. The living experience stimulates the sense for pleasant emotions or/and pass knowledge improving skills and abilities. Design elements' peculiarity: multifunction, transformable, friendly, enlightened, bright colors, unexpected graphic, smart materials, different kind of finishing, multimedia elements.

In conclusion the research suggests that all the elements of the kindergarten environment, must be both reassuring and educational in order to design a space "children centered" able to involve them in emotional and didactic experiences. In this way the experience at preschool becomes a pleasant training for the future life.

6 Conclusions

The results of the present work underline the importance for children to live in environment designed around their real attitudes, emotion and abilities. This work has identified a set of good practices and open rules to design pre-school environment, furniture, products and graphics according to children and educators "needs" with the final aim to improve children educational experience based on learning by doing.

Besides, the research highlights that all the material and graphic elements of children environment determine their living experience. The more these elements result familiar and reassuring, the more the experience will be pleasant influencing their security and wellness and as a consequence facilitating the educational experiences and development of their skills.

In summary the results of the research points out that reassuring elements and experiential ones have a decisive influence on the life's quality inside the kindergarten. Children need familiar environment and everyday object, in order to feel at ease living educative experiences in freedom.

For the future we hope to share the result of the research in order to apply and to develop the proposed open rules for designing not only preschool but suggesting a new design approach "children-centered".

References

1. Montessori, M.: Educazione per un mondo nuovo. Garzanti, Milano (1970)
2. Filippini, T., Vecchi, V., Malaguzzi, L.: The Hundred Languages of Children. Reggio Children Editore, Reggio Emilia (1996–2005)
3. Wise, B.K., Wise, J.A.: Children's human factors in the design of a pre-school educational furnishings system. In: Human Factors and Ergonomics Society Annual Meeting Proceedings, vol. 35, no. 8, pp. 541–545 (1991)
4. Cox, T., Griffiths, A., Rial-Gonzalez, E.: Work-related stress. Office for Official Publications of the European Communities, Luxembourg (2000)
5. Montessori, M.: Dall'infanzia all'adolescenza. Garzanti, Milano (1949)
6. Ministero dell'istruzione dell'Università e della Ricerca. Indicazioni nazionali per il curricolo della scuola dell'infanzia e del primo ciclo di istruzione, Italia (2012)
7. Edwards, C.P., Gandini, L., Forman, G.E.: The Hundred Languages of Children: The Reggio Emilia Approach to Early Childhood Education. Ablex Publishing Corporation, Norwood (1993)
8. Clark, A.: Ways of seeing: using the Mosaic approach to listen to young children's perspectives. In: Clark, A., Kjørholt, A.T., Moss, P. (eds.) Beyond Listening, pp. 12–28 (2005)
9. Druin, A.: The role of children in the design of new technology. Behav. Inf. Technol. **21**, 1–25 (2002)
10. Iversen, O.S., Smith, R.C., Dindler, C.: Child as protagonist: expanding the role of children in participatory design. In: Proceedings of the 2017 Conference on Interaction Design and Children, Stanford, California (2017)
11. Van Doorn, F., Stappers, P.J., Gielen, M.: Children as coresearchers: more than just a roleplay. In: IDC 2014 (2014)
12. Piaget, J.: Lo sviluppo mentale del bambino e altri studi di psicologia. Einaudi, Torino (1967)

Viral Marketing in Political Communication: Case Study of John Tsang's Campaign in the 2017 Hong Kong Chief Executive Election

Daren Chun-kit Poon and Sunny Sui-kwong Lam(✉)

The Open University of Hong Kong, Hong Kong, China
darenpck@gmail.com, ssklam@ouhk.edu.hk

Abstract. Viral marketing is one of the trends of the marketing strategy for building up a brand image because of the rapid development of internet. Viral marketing helps the information spreading quickly and widely on the internet through the sharing from the youngsters. Some politicians use the strategies of viral marketing to create noises and exposures for their political campaign and communication in Hong Kong through social media such as Facebook. This study will focus on how viral marketing works in political campaign and communication. A case study of John Tsang's campaign in the 2017 Hong Kong Chief Executive Election using the Berger's STEPPS framework is analyzed to assess how viral marketing was applied to political communication. The focus of analysis will be on the changes of the youth perceptions and attitudes toward John Tsang's image during the Umbrella Movement in 2014 and after his Election campaign in 2017.

Keywords: Viral marketing · Political communication · Social currency · Trigger · Emotion · Practical value · Social media

1 Introduction

Viral marketing is one of the new trends of the marketing strategy for building up a brand image. Viral marketing is a process that "one person gets infected and then their friend gets infected and then a friend of their friend gets infected, and so on" and this process is called multistep diffusion [1, p. 15]. It can gain a huge and positive impact, especially on the young generation. With the rapid global development of internet and Web 2.0, social media has become one of the main channels that information is spreading. According to Digital News Report 2016 by the Reuters Institute for the Study of Journalism, 51% of the respondents said they used social media as a source of news and 12% of the respondents said they used social media as the main source of news [2]. Furthermore, Facebook has become the major social media in Hong Kong, which provides the most connectivity and popularity to the public. Facebook is also "the most popular online social networking site among university students" [3, p. 1337]. It is a suitable and important platform for this study, which focuses on the youth generation in Hong Kong. The investigation will mainly focus on the viral

A. G. Ho (Ed.): AHFE 2019, AISC 974, pp. 151–161, 2020.
https://doi.org/10.1007/978-3-030-20500-3_16

marketing in Facebook, and that in other social media will be supplementary. This research aims to explore the strategies of viral marketing in political campaign and communication.

John Tsang's electoral campaign during the 2017 Hong Kong Chief Executive Election is a case study to discern how viral marketing was applied to promote John Tsang in the competitive election. In order to measure perceptions and emotional results, this study will compare the attitudes of the youngsters toward John Tsang during the Umbrella Movement and after the Election campaign. Generally, the model of viral marketing is used to study business cases in Hong Kong. It is a relatively new perspective to study how the viral marketing techniques were used in politics and how they brought emotional affections to young audiences about John Tsang's image in the Election.

The purpose of the study is to explore how the viral marketing was applied to political campaign and communication to promote the image of the candidate. The case study will use the Berger's STEPPS (social currency, triggers, emotion, practical value, public and stories) [4] framework to assess how John Tsang's campaign in the Election was going viral. In the meantime, the study will show the differences of the personal attitudes or thoughts of the youngsters in Hong Kong toward John Tsang during the Umbrella Movement and after the Election campaign.

2 Literature Reviews

Petrescu and Korgaonkar (2011, p. 218) mentioned that "viral marketing represents online and offline marketing activities performed to influence consumers to pass along commercial messages to other consumers" [5]. This briefly explains the working procedure of online marketing and offline marketing. Indeed, online marketing is using eWOM (electronic word of mouth), which means electronic consumer-to-consumer communication to spread the messages on social media. On the other hand, offline marketing is using different mass media such as television commercials, news and editorials to create the "Buzz" which means peer-to-peer communication in the society. Moreover, they also mentioned "its purpose includes the forwarding of business-generated commercial messages, through the Internet, from businesses to consumers and then to other consumers." This is the aim of viral marketing to deliver the messages from the generators and to spread the messages by the public. Therefore, it can gain the huge results with little effort [5].

McNair (2018, p. 4) mentions there are three principles of political communication as follows: (1) "All forms of communication undertaken by politicians and other political actors for the purpose of achieving specific objectives." (2) "Communication addressed to these actors by non-politicians such as voters and activists." (3) "Communication about these actors and their activities, as contained in news reports, editorials and other forms of media discussion of politics, such as blogs and social media posts" [6]. The case of John Tsang's electoral campaign has fulfilled these three principles. Firstly, John Tsang undertook all forms of communication including social media and mass media for the purpose of achieving specific objectives such as rising the voters or supporters and winning the 2017 Hong Kong Chief Executive Election.

Secondly, non-politicians such as the teenagers in Hong Kong addressed the communication to the actor John Tsang. Thirdly, the communication by John Tsang's campaign about him and his campaign activities, as contained in news reports, editorials and other forms of media discussion of politics, was mainly disseminated to the public by digital media such as blogs and social media posts.

In addition, there are three main perspectives to maintain the political communication including political organizations, media and citizens. One the one hand, the political organizations such as parties, government and public organizations may execute their political communication with different ways such as campaign, programme, advertising and public relation through the different media such as social media, press and broadcast online to reach the citizens. On the other hand, the citizens may also give some feedbacks such as opinion polls, letters, blogs and use-generated content (UGC) to the political organizations through the social media. Social media is playing a significant role to connect both the political organizations and the citizens. Meanwhile, McNair mentions the verbal or written statement and visual means of signification such as dress, make-up, and hairstyle, and logo design help to construct the political image or identity [6].

The Berger's STEPPS framework including social currency, triggers, emotion, practical value, public, and stories will be used to assess how a campaign goes viral.

Social currency refers to the people strive to look great, intelligent, or knowledgeable in front of others. If the content of the message will help the people look great, intelligent of knowledgeable, they will more likely share the message with others [4]. In the case of John Tsang's electoral campaign in 2017, his campaign crews used a lot of positive information to make people feeling good after they shared the messages such as the video "森美 X 曾俊華 拍住上!" (Sammy Leung X John Tsang Go Together!) [7].

Trigger is the motivation that drives individuals to take action on their latent readiness. When people are doing a decision-making, they are being in an unconscious state to take an intuitive response to cues in the environment. People will take action depending on the triggers of what they see, hear, read, or feel. Those people may have the specific intention before receiving the triggers. The triggers are being the cues to drive the people take the action by following their intention [4]. For instance, John Tsang invited a lot of famous people to shoot a video and make the series of video entitled "拍住上" (Go Together). People could easily remember the campaign with the famous peoples' video clips on social media.

Emotion is another factor to decide a message to be sharing or not. It is a kind of motivation to drive people share the message. However, not all the emotions will create the same effect on the reaction of the receivers. According to Berger's framework, the message of the high-arousal emotions is more likely to be sharing. On the contrary, the motivation of the sharing action will decrease if the message is of the low-arousal emotions [4]. For instance, there were some videos talking about how John Tsang supported the Hong Kong athletes such as the videos "同馬拉松選手加油!" (Encourage the Marathon Runners!) [8], "支持Rex爭取21連勝!" (Support Rex for 21 Consecutive Victories!) [9] and "到灣仔運動場爲學界田徑比賽打氣" (Going to the Wan Chai Sports Ground to Support the Inter-School Athletics Competition) [10] and so forth in John Tsang's electoral campaign. They were touching for the Hong Kong youngsters and the young athletes with the high-arousal emotions to attract more young supporters to his electoral campaign.

Practical value refers to the practical or useful information in the message. Berger finds out that people like to share the practical and useful information to others for helping others to solve problems. Practical value is truthful for the people and it helps the readers to understand a topic [4]. For instance, in John Tsang's electoral campaign in 2017, he provided a series of video of "我係曾俊華,你問我答" (I am John Tsang, you ask and I answer) in his Facebook page to give the information about him or other information that people want to know. This provided a lot of the information to the Hong Kong people to know more about him or the situation of the Hong Kong politics.

Public refers to highly visible and easily reachable of a viral content. Therefore, it encourages the imitation and a herd mentality enhanced by the public visibility. When people do not know what to do, they look to other for social cues [4]. For instance, John Tsang selected Facebook to be the main social media platform for sharing his campaign content to the public because Facebook is the easiest way to show his campaign's information to the target audience and it is a highly visible media for the young Hongkongers.

Stories are the storytelling of the content or the campaign. A story is more likely to be shared than an advertisement. It is because it is perceived as more trustworthy. Using stories to establish the relationships can build up the loyalty between the message sender and the receiver [4]. For instance, John Tsang filmed some video clips that talked about the stories and histories in Hong Kong such as the video "宣傳片: 香港故事-修復時刻" (Tsang・Promotion Video: Hong Kong Stories - Repairing Time) [11] using stories to build up the relationships between Hong Kong people and him. Those stories were used to increasing the chance for the campaign's message to be sharing through the videos.

All of the factors above are used for assessing the messaging factors that influence eWOM in the nonprofit sector. The more elements the campaign can fulfil, the higher chance the campaign goes viral.

Rival and Walach (2009, p. 40) mentioned, "In the field of Politics the motivations depend mainly on the context of the message (funny video, link to a party website…) and the sender profile ('highly-connected' person, political militant)." People should consider the emotions more than the practical value when they decide to share a political campaign or not. In addition, the high-arousal emotional messages are much more possible to be shared than the low-arousal emotional messages since people would, more likely, to share the content with the positive and affirmative to strive people to share the happiness in life to others. Thereafter, social media, broadcasting video and emailing campaigns are the usual ways to promote the candidates in politics [12].

According to Vernallis (2011, p. 74), "88% of all voters went online for political information" in the 2008 Presidential Election of America [13]. Within the high-tech generation, politics that has become technological is a big trend in the world. Most of the youngsters who are interested in political issues are willing to go online for searching more information about the issues. Therefore, social media is a great media platform to let the youngsters receive and spread the information they want. Vernallis also mentions (2011, p. 74) "audiovisually rich political clips were forwarded via Facebook (Obama had 2.5 million subscribers), nested in blogs, marked as 'favorites' on YouTube, and both created for and downloaded from the My Barack Obama website (MyBo). It is widely acknowledged that Obama won the election, at least in part, due to his MyBo site and his appeal to youth" [13]. The Obama's crews generated and delivered the message through the social media in video format to target the youth voters. Even the American President candidate was willing to use social media to reach the youngsters with a high level of effectiveness and efficiency. This shows social media is playing a significant role in political campaign among the new generation.

As Kwong (2015, p. 280) mentioned, "For most locals and international media, the Umbrella Movement was a student-led campaign for 'an election method for the Chief Executive that ensures their rights of choosing'." "Even worse, the People's Daily linked the Occupy movement to seeking Hong Kong's 'self-determination' and even 'independence'." They tell us that the different stances of the locals and international media to report the Umbrella Movement [14]. The China mainstream news media may choose a pro-communist stance when they are doing the news report about the Umbrella Movement by using some extreme words to identify the Movement such as "self-determination" or "independence". Therefore, most of the Hong Kong people choose to use internet media such as social media to search for information about the Umbrella Movement. "Disappointment with mainstream media resulted in the rise of online media." Therefore, "Internet media users believed that mainstream media were taking a pro-government stance while online media presented the truth within the occupy sites" [14, p. 287]. People are more likely to share and receive the information about the political campaign on the internet since there are many perspectives they can choose. Instead of receiving the information from the mainstream news media, people are more likely to receive and trust the information that come from different lobbying organizations, social media users and the online news media on the internet.

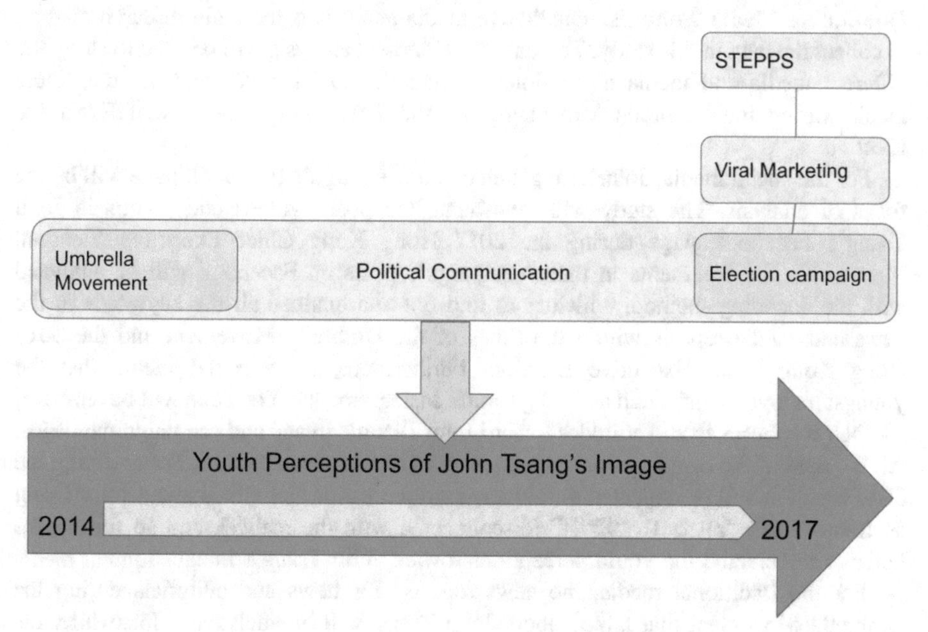

Fig. 1. Changing perceptions of John Tsang's image by youth generation

3 Theoretical Framework

Figure 1 is a timeline to represent the change of the youth attitudes and perceptions toward John Tsang's image. It is influenced by the political communication such as campaign, advertising, program delivering through the social media and the traditional media from Umbrella Movement in 2014 to his Election campaign in 2017. The perceptions changed because of the viral marketing strategy of STEPPS used in the 2017 Hong Kong Chief Executive Election campaign. Based on this theoretical framework, three research questions are developed to study how John Tsang and his campaign crews applied viral marketing strategies to political communication to affect the youngsters through social media. In addition, qualitative media analysis and focus group are used to collecting data for empirical studies.

3.1 Research Questions

1. What were the strategies and viral marketing techniques employed to make the John Tsang's Election campaign go viral?
2. How did the viral marketing apply to political communication to promote John Tsang's image?
3. How did the viral marketing influence and change youth attitudes and perceptions toward the image of John Tsang after the Election?

3.2 Research Method and Methodology

Qualitative Media Analysis. Qualitative media analysis is the main research method to collect the data in this study. Textual and document analysis will be used to study the different media and media discussions through the social media and the traditional media during the Umbrella Movement and the 2017 Hong Kong Chief Executive Election.

For the social media, John Tsang's electoral campaign's Facebook page will be the focus of analysis. The study will investigate the posts, videos and photos in John Tsang's Facebook page during the 2017 Hong Kong Chief Executive Election. Meanwhile, the comments in those campaign's posts on Facebook will be analyzed with the decoding method, which is to find out and analyze all the keywords in the news and media reports within the times of the Umbrella Movement and the 2017 Hong Kong Chief Executive Election. Furthermore, those social media that the youngsters love to visit such as Lihkg forum, Instagram, and YouTube will be analyzed for their responses to and attitudes toward John Tsang's image and campaign messages. All the data of the campaign posts, photos and videos related to John Tsang during the 2017 Election will be collected to find any extreme words that talked about John Tsang such as "hate", "respect", "離地" (disconnected with the reality), and so forth. This helps to understand the youth perceptions toward John Tsang's image.

For the traditional media, the news reports, TV news and editorials during the Umbrella Movement that talked about John Tsang will be analyzed. Meanwhile, the news after the 2017 Election will also be studied for comparing the changing perceptions and attitudes toward John Tsang. The traditional media will be divided into two political stances such as the neutral political stance and the "Pro-communist" or "Leftist" stance, as we can see the different perceptions and attitudes toward John Tsang by the two different political stances of news media. It will also use the decoding method to find out all the extreme words used in the news reports, TV news and editorials. For instance, there are two news reports from the different news media that are Apple Daily and Etnet, and one from a "Pro-communist" news media, Ta Kung Pao. The two news reports from Apple Daily and Etnet used some kind of neutral words to elaborate John Tsang. On the contrary, the article of Ta Kung Pao "羅馬非一天建成 摧毀只一念之差 財爺憂法治傷害更甚經濟" (Rome is not built in one day. The foundation of Hong Kong is destroyed by a wrong decision. John Tsang worries about damages in the rule of law more than the economy) used some extreme and negative words such as "摧毀" (destroy), "憂" (worry) and "傷害" (damage) to exaggerate the negative level of the contents [15].

Focus Group. Focus group is another research method in this study. There will be 5 people in a group and 4 groups in total. The target people of the focus group are the Hong Kong youngsters who age 20–27. This age range to be the focus group is because they are most likely the participants of the Umbrella Movement and have the strong impression and critical thinking toward the political issues in the society. Therefore, 20–27 years old focus group participants should have the strong impression and critical thinking to comment on the Umbrella Movement and the perceptions toward John Tsang. For the political status, 10 of the interviewees who supported the Umbrella Movement will be in two focus groups and other 10 of the interviewees who did not support the Umbrella Movement will be in another two groups. The time of the focus group interview should be around one to two hours. I will let the focus group participants have casual conversation about the political issues for 15 min to warm up and to generate a rapport. After that, they may take about an hour and a half to focus on John Tsang's political campaign and his image during the Umbrella Movement and after the Election campaign in 2017.

The purpose of the focus group is to discuss John Tsang's images that the local youngsters have perceived during Umbrella Movement and John Tsang's political campaign of the 2017 Hong Kong Chief Executive Election. Last but not the least, they will be encouraged to discuss any changes of the feelings and emotions toward John Tsang's images after the political campaign in the 2017 Hong Kong Chief Executive Election.

4 Expected Results and Findings

For the expected results and findings, three major points are expected to be found by the research. Firstly, it is expected to know how John Tsang's electoral campaign has satisfied the principles of STEPPS and made it go viral. Secondly, the strategies that John Tsang's campaign crews used to promote John Tsang's image should be identified. Last but not the least, the youth attitudes and perceptions toward John Tsang's image after the Election campaign should be promoted and changed to be more positive.

4.1 Data Analysis and Results

STEPPS Principles (RQ1 + RQ2). For the social currency, John Tsang's Facebook page has uploaded a video "森美 X 曾俊華 拍住上!" (Sammy Leung X John Tsang Go Together!) [7], in which a famous Hong Kong artist, Sammy Leung, was invited to talk about the experiences when he was getting along with John Tsang. He shared his interesting memories about John Tsang that people may want to know in the video. Thus, people were willing to share the video with others for its interesting content about John Tsang, and there were about 566,000 viewers for the video. In the video, Sammy Leung talked about how John Tsang was acting as a fencing coach. He was a caring coach who always encouraged and taught in accordance with the student's different aptitudes. This revealed his positive image of flexibility and caring about others to the Hong Kong people.

For the triggers, John Tsang invited a number of celebrities in Hong Kong to support and promote him. In addition to the video "Sammy Leung X John Tsang Go Together!" the Election campaign created another video "杜sir飯局" (Meal with Johnnie To) [16] in John Tsang's Facebook page. Many local celebrities such as Johnnie To, who is a famous Hong Kong film director and producer, Benjamin Au, who is a host in Television Broadcast Limited, and James Tien Pei-chun, who is a politician in Hong Kong, were invited to have a meal and discuss about the politics in Hong Kong with John Tsang. Those celebrities were having high exposure rates in television or some broadcast radio shows, and people may associate to John Tsang when they saw or listened to those celebrities. There were around 52,000 viewers for this video in Facebook.

For the emotions, John Tsang's Facebook page has uploaded a video "支持Rex爭取21連勝!" (Support Rex for 21 Consecutive Victories!) [9]. In the video, John Tsang was having a great interaction with Rex Tso, who is a famous super-flyweight boxer of Hong Kong. This revealed his supporting attitude toward the young Hong Kong athlete with action. Giving the inspiration to the young Hong Kong athlete and inspiring the Hong Kong youngsters, this helped to create the high-arousal emotions with the positive associations to the Hong Kong youngsters. Therefore, they were willing to share the video, and there were about 127,000 viewers for this video. John Tsang used this video to build up a positive image among the Hong Kong youngsters by the message that he cared about the Hong Kong sports development and the athletes to gain their supports.

For the practical value, there was a video "我係曾俊華,你問我答" (I am John Tsang, you ask and I answer) [17] in John Tsang's Facebook page. In the video, John Tsang was answering the questions from the citizens with the practical and useful information based on his own experience and knowledge. For example, there was a citizen, Mary Ho, asking John Tsang about the idea of immigration. John Tsang could explain the situation in Hong Kong and give a rational suggestion for the citizen in order to provide the practical value and advice to help her for the question. Meanwhile, he also talked about his own experience in immigrating to give his own tips and advice truthfully. Thus, people were willing to share his video since it provided useful and truthful information, and there were about 143,000 viewers. John Tsang used this video to build up the positive communication with the youngsters in Hong Kong through answering the questions and giving the valuable suggestions. This video helped to create a positive image of John Tsang, who was willing to listen to and communicate with the youngsters in Hong Kong.

For public, there were a series of video "曾俊華落區" (John Tsang Visits the Local Community) such as "曾俊華落區:大圍" (John Tsang Visits the Local Community: Tai Wai) [18] and "曾俊華落區:美孚" (John Tsang Visits the Local Community: Mei Foo) [19]. John Tsang was going out to the community and interacting with the citizens to increase his public visibility. Moreover, he distributed his political platform and some of the promotion materials such as a medal to promote his electoral campaign crews. By means of the face-to-face interactions between John Tsang and the citizens, his image was becoming more trustful and positive. There were respectively 260,000 viewers for the video "John Tsang Visits the Local Community: Tai Wai" and 356,000 viewers for the video "John Tsang Visits the Local Community: Mei Foo". In these videos, John Tsang created a positive image that he was willing to have face-to-face communication with Hong Kong people. This revealed he was down-to-earth and proactive with his political communication and idea for Hong Kong and the Hong Kong youngsters.

For the stories, John Tsang posted the videos "宣傳片: 香港故事-修復時刻" (Tsang・Promotion Video: Hong Kong Stories - Repairing Time) [11] to share the ten real stories of Hong Kong people to the public. One of the stories "香港故事-修復時刻: 被偷走的記憶" (Hong Kong Stories - Repairing Time: Stolen Memory) [20] was about two young Hong Kong girls to take care of their old grandmother who has Alzheimer disease. John Tsang was spending some time to be along with them in the video, listening to the difficulties of the young girls and their old grandmother. Furthermore, John Tsang told the story about him and his wife to the audience to share his own story in order to let people feel that he was trustworthy. Meanwhile, he was spreading his political concept that was caring about all of the Hong Kong people when people were watching those stories. The young Hongkongers may gain the high-arousal emotions after watching the videos and they were willing to share them to their friends. There were about 54,000 viewers for this video. John Tsang created a positive image of caring about the groups of weaknesses in Hong Kong and showed his willingness to help different people with different difficulties in Hong Kong.

5 Conclusion and Discussion

In conclusion, viral marketing is actually playing a major role in the political communication of John Tsang's campaign in the 2017 Hong Kong Chief Executive Election. It was helping him to build up the positive image and to change the Hong Kong youngsters' attitudes and perceptions toward his image using those STEPPS strategies of viral marketing on the social media. The videos and contents in John Tsang's Facebook page helped to make his campaign go viral on the internet. Meanwhile, it is important that the viral marketing applied to John Tsang's political communication through not only the social media, but also the traditional media. It was because those traditional media were still being the big and influential communication channels for Hong Kong people to receive information in the society. Thus, John Tsang and his campaign crews created his positive image successfully through both social media and traditional media. However, a critical limitation was that the youngsters in Hong Kong have no right to vote in the Election. Thus, it was difficult to measure how success was the viral marketing of his campaign and political communication to a certain extent.

References

1. Fitzgerald, M.: The myth about viral marketing. MIT Sloan Manag. Rev. **54**(3), 15 (2013)
2. Newman, N., Fletcher, R., Levy, D.A.L., Nielsen, R.K.: Reuters Institute Digital News Report 2016. Reuters Institute for the Study of Journalism. University of Oxford, UK (2016)
3. Cheung, C.M.K., Chiu, P.Y., Lee, M.K.O.: Online social networks: why do students use facebook? Comput. Hum. Behav. **27**(4), 1337–1343 (2011)
4. Berger, J.: Contagious: Why Things Catch On. Simon & Schuster, New York (2013)
5. Petrescu, M., Korgaonkar, P.: Viral advertising: definitional review and synthesis. J. Internet Commer. **10**(3), 208–226 (2011)
6. McNair, B.: An Introduction to Political Communication. Routledge, London (2018)
7. [John Tsang 曾俊華].:【森美 X 曾俊華 拍住上!】(Sammy Leung X John Tsang Go Together!) [Video File]. https://www.facebook.com/johntsangpage/videos/1842254919380485/. Accessed 28 Feb 2019
8. [John Tsang 曾俊華].:【同馬拉松選手加油!】(Encourage the Marathon Runners!) [Video File]. https://www.facebook.com/johntsangpage/videos/1853046124968031/. Accessed 28 Feb 2019
9. [John Tsang 曾俊華].:【支持Rex爭取21連勝!】(Support Rex for 21 Consecutive Victories!) [Video File]. https://www.facebook.com/johntsangpage/videos/1864493713823272/. Accessed 28 Feb 2019
10. [John Tsang 曾俊華].:【到灣仔運動場爲學界田徑比賽打氣】(Going to the Wan Chai Sports Ground to Support the Inter-School Athletics Competition) [Video File]. https://www.facebook.com/johntsangpage/videos/1863005263972117/. Accessed 28 Feb 2019
11. [John Tsang 曾俊華].:【宣傳片】香港故事-修復時刻 (Tsang • Promotion Video: Hong Kong Stories - Repairing Time) [Video File]. https://www.facebook.com/johntsangpage/videos/272842873239254/. Accessed 28 Feb 2019
12. Rival, J.B., Walach, J.: The use of viral marketing in politics: a case study of the 2007 French presidential election. [Bachelor's Thesis within Business Administration]. Jönköping International Business School, Jönköping University (2009). http://www.diva-portal.org/smash/get/diva2:225806/FULLTEXT01.pdf. Accessed 1 Feb 2019
13. Vernallis, C.: Audiovisual change: viral web media and the Obama campaign. Cine. J. **50**(4), 73–97 (2011)
14. Kwong, Y.H.: The dynamics of mainstream and internet alternative media in Hong Kong: a case study of the Umbrella Movement. Int. J. China Stud. **6**(3), 273–295 (2015)
15. Cheung, T.S.: 羅馬非一天建成 摧毀只一念之差 財爺憂法治傷害更甚經濟 (Rome is not built in one day. The foundation of Hong Kong is destroyed by a wrong decision. John Tsang worries about damages in the rule of law more than the economy). Tai Kung Pao, p. A02 (2014)

16. [John Tsang 曾俊華].:【杜sir飯局】(Meal with Johnnie To) [Video File]. https://www.facebook.com/johntsangpage/videos/1874462422826401/. Accessed 28 Feb 2019
17. [John Tsang 曾俊華].:【我係曾俊華,你問我答】(I am John Tsang, you ask and I answer) [Video File]. https://www.facebook.com/johntsangpage/videos/1845292129076764/. Accessed 28 Feb 2019
18. [John Tsang 曾俊華].:【曾俊華落區:大圍】(John Tsang Visits the Local Community: Tai Wai) [Video File]. https://www.facebook.com/johntsangpage/videos/1871777683094875/. Accessed 28 Feb 2019
19. [John Tsang 曾俊華].:【曾俊華落區:美孚】(John Tsang Visits the Local Community: Mei Foo) [Video File]. https://www.facebook.com/johntsangpage/videos/1873476816258295/. Accessed 28 Feb 2019
20. [John Tsang 曾俊華].:【香港故事-修復時刻: 被偷走的記憶】(Hong Kong Stories - Repairing Time: Stolen Memory) [Video File]. https://www.facebook.com/johntsangpage/videos/283715335485341/. Accessed 28 Feb 2019

Study on How Television Commercials Affect Consumer Reactions with Visual Strategies

Pikki Fung(✉) and Amic G. Ho

Department of Creative Arts, The Open University of Hong Kong, Jocly Club Campus, Ho Man Tin, Hong Kong
fpk0714@gmail.com, amicgho@ouhk.edu.hk

Abstract. Visual communication is not only significant and indispensable in outstanding in advertising, but also a form of language for art to visually communicate the idea of advertisements. This study aimed at discovering how can visual elements be applied in television commercials to associate with the observers' emotion psychologically as well as to affect their decision-making process behaviourally through conducting the depth interview, content analysis and library research. This is done by showing the application of visual elements and the findings on the research in term of lighting and colours, mood and tones, signs, symbols, key signifiers, and typeface. This research focuses on the use of visual strategies, and explore how they could evoke male and female audience using different approaches. Another goal of the study is to reveal the impacts in terms of sociology and psychology when certain visual elements applied in television commercials.

Keywords: Visual strategy · Visual communication · Emotion · Typography · Colour usage · Semiotics

1 Introduction

Advertising plays a powerful and indispensable role to arouse the desires to the audience. The consumers ought to make the decision of what to choose among the countless products and services. The easiest and most direct way to stand out from the competitors is to advertise. Throughout these few decades, television is a dominant media in advertising. "Television is the pre-eminent brand awareness channel…Television will remain the principal display medium for many years to come" [1]. In order to enlarge the market share within the industry, advertisers tend to expend a high ratio of advertising budgets on television communication. As a result, it is worthwhile analysing the television commercial.

No matter for the male or female, their preferences on products can be totally different, but there are some common visual preferences for both the male and female. Advertising designers are responsible for constructing the advertising which can easily grab the attention of the target audience and arouse their desires and reactions eventually. Through the observation and analysis of the perception of the audience, some significant visual strategies can be defined successfully for the use of advertising design.

A. G. Ho (Ed.): AHFE 2019, AISC 974, pp. 162–173, 2020.
https://doi.org/10.1007/978-3-030-20500-3_17

For the video-context analysis, this study mainly focuses on three of the visual indicators. The first indicator is colour perception and association which refers to the colour usages, lighting, mood and tone of the TVC. The second indicator is typeface which refers to the type design of the TVC. The third indicator is signed which refers to the objects shown in TVC.

To analyse the effectiveness of visual strategies in TVC which refers to how do the visual strategies employed in TVC to evoke the emotional impacts and consequently affect the consumer behaviours, three of the research questions are proposed for processing the following research sections:

RQ1. How advertising designers apply the visual strategies in television commercials and reinforce the core message to the specific target?
RQ2. Can visual strategies be used to arouse the interest of purchase or speed up the decision-making process?
RQ3. Are there any impacts or implicit messages when certain visual elements employed in television commercials?

2 Aims and Objectives

This study focuses on the visual analysis of TVC with three aims. First, this study is to analyse the usage of visual strategies used in TVC and explore how they could grab the attention between male and female audience in different visual approaches. The visual components applied in TVC which is chosen as samples will be described in detail. Second, this study is to discover how do the visual elements applied in TVC to associate with the audience' emotion psychologically and affect their decision-making process behaviourally. The respondents will express their ideal TVC about visual associations, feelings and visual preferences. Third, in the last section of this study, the impacts in terms of sociology and psychology will be revealed in the situation of applying certain visual elements on TVC.

3 Literature Review

3.1 Adoption of Visual Strategies in TVC

The adoption of suitable visual strategies is important for advertising. For the advertisers, the appropriate use of visual strategies affects the buying desires and decision-making process positively. "Color can contribute to differentiating products from competitors and creating positive or negative feelings about products" [2].

3.2 The Difference Between Visual Perception of Male and Female

Hence, some researches have been shown that the visual perception between male and female can be certainly different. Due to human history, some visual strategies are more attractive to male or female specifically. "Researchers have suggested that colour associations may have been formulated early in human history" [2]. A social role is another reason for male and female perceiving TVC in different ways. "Once a child accepts membership in a gender group, he or she comes to value and adopt the social role associated with their gender label and this gender role [3]. "The social role of early females (i.e., foraging for plant food and caretaking of infants) may have evolved in girls compared to boys a greater specialization for color processing and a greater preference for objects with a pink or reddish color" [3].

3.3 Theories

To explain human behaviours clearly and logically, several theories are important in the related phenomenon. One of the theories is *Cognitive behavioural theory* which was proposed by Dr. Aaron T. Beck. "It is based on the idea that how we think (cognition), how we feel (emotion) and how we act (behaviour) all interact together" [4, 5]. Robert J Lavidge and Gary A Steiner developed *Hierarchy of effects theory* which is closely related to consumer behaviour and commonly used in researches. The model includes seven stages in a linear way. The hierarchy represents the progression of learning and decision-making consumer experiences as a result of advertising [6]. Another framework *Information theory* which was defined by Claude E. Shannon to elaborate some noise is created among the process of delivering and receiving the messages.

4 Research Methods

4.1 Nature of TVC

In order to prevent any bias and preferences towards the brands and products and ensure that the decision of the audience is entirely based on the visual factors, the TVCs which are selected for analysing should comply with the following standards strictly:

- The content of TVCs is not mainly targeted for the respondents.
- The TVCs are widely broadcasted in the origin but not the place the respondents live in.
- Three of the visual strategies aimed to analyse can be perceived obviously in each of the TVCs.

On the other hands, the TVCs which are selected to analyse advertised between 2013 and 2018 in this study as it is more accurate when analysing and elaborating the symbolic meaning.

The TVCs are searched through YouTube and some advertising Facebook pages. To investigate the visual preference of the female audience, the fashion TVCs from Japan are selected which are the about beauty service and clothes sales. For investigating the visual preference of the male audience, the shampoo TVCs are selected for researching which are specially made for men to use. For understanding how do both male and female audience perceive the identical elements, the TVCs of neutral products are selected as well which the airlines are advertising.

4.2 Content Analysis for Semantic Knowledge

Berger [7] analysed the signs and their functions in different types of media communications such as magazine advertisements and films. "Content analysis is a research technique for the systematic classification and description of communication content according to certain usually predetermined categories. It may involve quantitative or qualitative analysis or both". Through conducting the content analysis, the impacts and values can be found in the material. Hoogs, A., Rittscher, J., Stein, G., &. Schmiederer, J. searched for the semantic knowledge and meanings through WordNet which "is a lexical ontology that attempts to organize lexical information hierarchically in terms of word meanings… Our results indicate that using WordNet can substantially improve the accuracy and specificity of video annotations over using visual processing alone." [8]. As semiotics are frequently used in almost all of the TVC to convey some complicated meanings with the signs and texts, the meanings and functions of the signs in TVC will be referred in the study through WordNet in order to research the impacts caused or implicit messages created when the certain visual elements employed in TVC.

4.3 Demonstration of TVC

Some scholars analysed the video contents and some of them tried to clip the video as the still images. Steve, P. applied 'dragging' in his research to analyse tourism TV commercials. "The term dragging refers to associating elements from separate files of representation to frame an event and/or create a new value" [9]. According to Steve's research, 33 still frames from the TVC were selected to use for the visuals recall test in research. Following with the visuals recall test, he analysed the association among the still photos.

Moreover, the textual description will also be used in this study so the detail can be easily and clearly to point out.

The software ImToo is used to convert the video the still photos. As six of the TVC aimed to study is within 30 s, so the setting of clipping the frames is the same. To show each scene clearly, 30 frames were captured averagely in each TVC. After that, the apps PhotoScape X is used to combine 30 frames in the whole one.

4.4 In-Depth Interview

It is one of the qualitative researches and some meaningful primary data can be easily collected through the interview. In this part of the research, interviewees will be investigated about the perception and preference of visual strategies in TVC. The total number of 63 interviewees will be included in this study. Interviewees who are at the age of 19 to 36 will be chosen as the interviewees in this study for several reasons.

The purposes of conducting interview are to find out which one of the visual elements impress the interviewees the most when they are viewing the TVC provided, examine and compare the male-oriented and female-oriented visual strategies and eventually come up with the ideal advertising that can arouse their awareness for male and female audience with the interviewees' supportive reasons.

4.5 Procedure

In the first stage, two sets of TVs within the same product category will be viewed by the interviewees and they need to choose which one of the TVC is more attractive for them. After that, they ought to express what visual element is the most impressive in the TVC which they choose. In the second stage, corresponding questions which are related to the visual elements they prefer will be asked to gain the important detail and supportive reasons. In the third stage, the interviewees will explain some impacts or implicit messages in TVC they have viewed openly (Fig. 1).

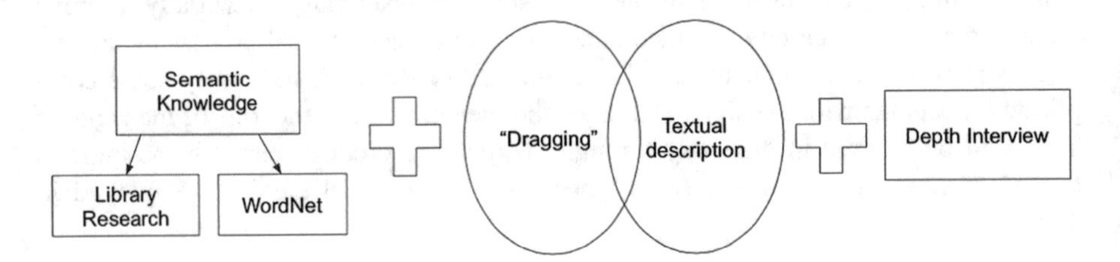

Fig. 1. Research methodology

5 Findings

5.1 The Different Visual Preference for TVC Between Male and Female

One of the purposes of this study is to find out how advertising designers apply the visual strategies in TVC and therefore reinforce the core message to the specific target. Since different visual preferences on visual strategy between male and female can, therefore, cause to some special visual strategies by the advertising designers for either of them, so the data of their visual preference will be analysed briefly first. Next, in order to access some useful information, the respondents were responsible to answer which advertisement is more attractive for them with advertising materials by category and they needed to explain what they could associate with so that it can ensure that the data collected is the most accurate.

Generally, it proves that male and female percept visual elements in a different way from the past until present. There are 50% of the male respondents put more attention on the signs used on TVC than the 43.3% of female respondents. Also, it reveals that colour association is the most important visual strategy for both male and female respondents because 58.8% of the respondents claimed that colour associations are the most impressive element. However, there is none of the male respondents and are only 6.1% of female respondents choose typeface as their most concern elements. This result shows that advertising designers can reinforce the core message to the target audience by employing visual strategies especially.

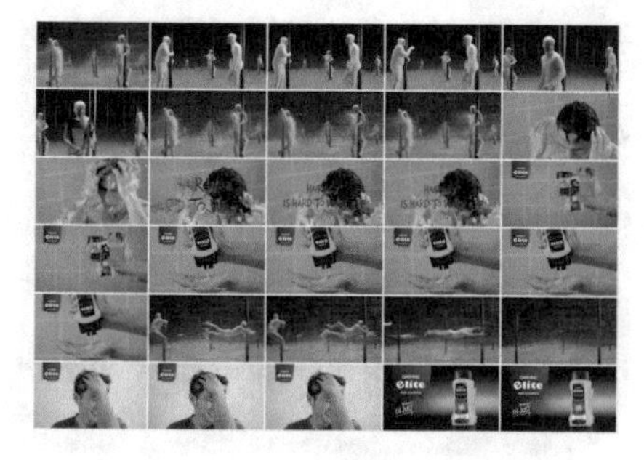

Fig. 2. Ad1 Dashing Elite: Sticky Pole Men

Fig. 3. Ad2 Dashing Elite: Sticky Pole Men

For more specific, in men's product category, the male respondents conveyed that Ad2 [10] is more attractive compared with Ad1 [11]. According to the result collected, 70% of the male respondents said that they preferred the Ad2 more than Ad1 instead. Besides, respondents chose Ad2 expressed their visual preference on TVC, 50% of them pay their attention to the signs on Ad2 and the rest of 50% male respondents pay their attention to the colour association on Ad2. However, none of the respondents said that typeface is attractive on TVC. Firstly, in term of the signs, most of them explained that the prairie shown in Ad2 was to emphasise its natural ingredients and fresh and comfortable feeling after using the shampoo. Also, the male respondents pointed out that the charming smile of the main actor in the party suggested that the consumers could be as confident as the main actor as long as they purchased the shampoo and used it everyday. They also noticed that the white laboratory and coat signified that Gerovital was the expert of shampoo and the product was developed through exact experiments. In term of the colour association, the male respondents who thought colour association is more attractive than the signs in Ad2 said that they realised the shampoo should be made of mild formula and users feel refreshing due to colour usage of the prairie in Lawn Green and the brown trunk on TVC. Moreover, the male respondents said that some psychedelic colors as if the disco ball reflects on the mirror imply the meaning of being trendy, modern and popular (Figs. 2, 3).

Fig. 4. Ad3 SHOPLIST: 色石篇 Megasale

Fig. 5. Ad4 銀座カラー: 「ミス・チョイスの罠」

For the women's product, the female respondents expressed that Ad3 [12] is more attractive compared with Ad4 (銀座カラー: 「ミス・チョイスの罠」) [13]. According to the result collected, 61.9% of the female respondents said that they preferred the Ad3 than Ad4. Moreover, respondents chose Ad3 showed their visual preference on Ad3 that 69.2% of them concern colour association most, 23.08% of them concern the signs used and only 7.7% of them focus on the typeface when they are watching the printed material of Ad3. Firstly, in term of the colour association, most of them explained that the colourful scenes like a rainbow, birthday cake and theme park imply that online shopping is full of happiness and there are different styles of clothing for selling in the whole year. Also, few of them mentioned that the bright colours such the pure white, shape yellow and baby pink colours were associated with the positive images and it can easily attract the female. Secondly, in term of the signs used on Ad3, the respondents said that different backdrops imply that the company offers many clothings for different contents. They also mentioned that the main actress got dressed in different styles convey that regardless of the age, female can dress up what they want as long as they enjoy it. Thirdly, in term of the typeface showed on Ad3, one of the respondents said that the typefaces were designed matching with the clothings of the actress and the scenes in colour which was so playful and they could associate with the different ages (Figs. 4, 5).

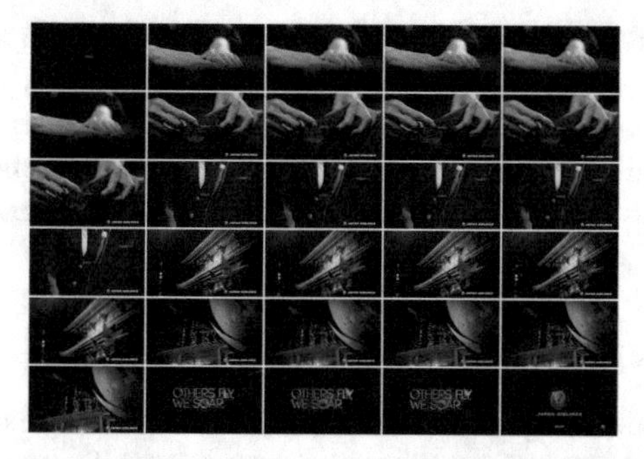

Fig. 6. Ad5 Japan Airlines: Perfectly Prepared

Fig. 7. Ad6 Vietnam Airlines: Reach Further - Proud to Be 4-star Airlines

For neutral product, both male and female were able to choose which advertising they enjoy watching more. The result is that 76.9% of the female respondents and 50% of the male respondents like watching Ad5 [14] instead of Ad6 [15]. For the female respondents, 90% of them thought that color association is more important when they are watching the TVC while only 10% of them chose the signs as the important visual strategy on TVC. On the other hand, 80% of the male respondents showed their preference for colour association and only 20% of the male respondents thought that the signs are important visual elements on TVC. In term of the colour associations from the female respondents, they tended to associate the dark red and black colours with the intangible value. One of the female respondents expressed that the colours were so mysterious and therefore she imagined Japan is mysterious and traditional as well. Another female respondent said that the black and red colours implied something fancy and professional. Meanwhile, the colour associations from the male respondents are more related to the emotion. They explained that the colours were full of passions and the consumers of Japan Airlines were pretty awesome (Figs. 6, 7).

To sum up, referring to data and the explanations from the respondents, it shows that there are different visual perceptions and preferences from male and female. So, if the advertising designers would like to further or reinforce the core message to their target audience, especially for either male or female, they can try to apply some unique visual strategies for the target audience. Moreover, designers can further develop some associations between the brandings and the visual presentations. In the interview, it is found that some important and interesting messages can be interpreted through one or even all of the visual strategies. So, if there are more connections between the visual presentation and advertisers, there are deeper impressions created for the target audience.

5.2 The Ideal Use of Visual Strategies on TVC

Another purpose of the study is to is to discover how do the visual elements applied in TVC to associate with the audience' emotion psychologically and affect their decision-making process behaviourally, therefore, the interviewees' ideal TVC about the visual association, feelings and visual preferences will be analysed. In the interview, over 97% of the respondents claimed that all visual strategies and visual elements applied to TVC will affect their decision-making process of purchase and emotions towards the branding. First of all, regarding the ideal use of colour association, most of the female respondents prefer some bright and eye-catching colours since they thought the strong contrast of colour was easy to grab the attention and stand out from other TVCs. While for the male respondents, most of them replied that they prefer a dark colour like black, dark blue and dark green colours. Surprisingly, both male and female respondents explained that they would love to see their favourite colours applied on TVCs.

Secondly, in term of the ideal use of the sign on TVC, both male and female respondents prefer watching the living creature on TVC, such as human beings and pets. For the most female respondents, they would like to watch the little kids and handsome men on TVC. While for the male respondents, they would enjoy watching the young ladies and some of them mentioned the luxury automobile on TVC. The most popular reason is that they want to be as attractive as the actor or actress on TVC which reveals that the advertising shows how the target audience wants to be and even communicate that how they should be.

5.3 The Positive and Negative Impacts of Applying Visual Strategies on TVC

Another reason for conducting this research is to find out the impacts when advertising designers applying certain visual elements on TVC. Some respondents mentioned that the visual strategies may create stereotype to the audience because some TVC is designed to target a specific gender or ethnic groups but there are always some stakeholders against the advertising message. A respondent took the beer advertising as an example, she said that the young ladies with scanty dress can surely attract the male audience who are mainly targeted by the advertiser but some parents worry that the sexy women on TVC may affect the value of teenagers.

Also, some audience who are in low self-esteem and always looks down on themselves, especially the female audience, they are easily affected by the signs portrayed on the TVC, such as the perfect body figure and great skin condition and then it causes to the materialistic problem. It actually shows the cognition, emotion and behaviours of the audience from the Cognitive behavioural theory.

Besides, although part of the key messages from the TVC can be received correctly by most of the audience, there are still some misleading ideas existed referring to the information theory. When thinking of the most unforgotten TVC, the respondents could generally answer with the implicit or symbolic meaning of the TVC. One of the male respondents took a chocolate TVC as an example, he said the young ladies were very proactive when meeting the men online. However, as male and female perceived the visual elements in different ways and many sources of noise from society, it finally may form the wrong messages that are not intended to communicate by the advertisers. Actually, the ladies from the chocolate TVC mentioned were just talking about which flavour of chocolate to eat.

When asking how the visual strategies affected their emotions and feelings, the respondents tended to give out very positive explanations. Most of the respondents believed that visual strategies such as the famous spokesman and special colour patterns employed on TVC enable them to associate with the brand colours and brand images. One of the middle-aged male respondents mentioned that it was so profound that his favourite cartoon characteristics from Dragon Ball in a financial TVC which amazed him and then later he had become the client of that advertiser. Along with watching the view, the particular atmosphere is created by the message and attitude of visual strategies and eventually form the stereotypes to the audience.

Moreover, some respondents explained that the visual strategies could sometimes show the atmosphere and intangible feelings of the products advertised, such as perfume and shampoo. It can successfully emphasise the product features and characteristics which are difficult to present without visual elements.

6 Limitations

Due to the lack of supports and resources, the number of interview participants is not too much and if there are more participants, the result and analysis can be more detail. Moreover, although the personal preferences had been concerned, the brand personality of the brands themselves still exist. For example, the respondents may overlook some special and unique design elements on the examples of TVC. Furthermore, the colour association may be too unilateral because it cannot be discussed individually without the brand personality.

7 Conclusion

To increase the effectiveness of advertising communication, the need for research on visual preferences between male and female especially on TVC is necessary which is seldom studied by scholars. This study was completed through depth interview, description by text and "Dragging" and acquiring semantic knowledge. It explored the visual preferences of male and female, their ideal TVC in term of visual strategies and the impacts of applying visual strategies on TVC. The important findings are shown as following:

- Male and female value the colour association as the most important visual strategy on TVC. The male would like to see some dark colours and female would like to see some bright and fancy colours.
- Male puts more attention on the signs than the female. The male prefers watching the young ladies, automobile and something awesome on TVC. While the female prefers watching the kids and handsome men on TVC.
- Visual strategies showcase the intangible value and atmosphere of the product and branding.
- The audience may be misled by the advertised TVC because of the noise and source.

References

1. Adspend forecasts March 2018 executive summary (n.d.). Retrieved Winter 2018. https://www.zenithmedia.com/wp-content/uploads/2018/03/Adspend-forecasts-March-2018-executive-summary.pdf
2. Grossman, R.P., Wisenblit, J.Z.: What we know about consumers' color choices. J. Mark. Pract.: Appl. Mark. Sci. **5**(3), 78–88 (1999). https://doi.org/10.1108/eum0000000004565
3. Alexander, G.M.: An evolutionary perspective of sex-typed toy preferences: pink, blue, and the brain (n.d.). Retrieved Winter 2018. https://link.springer.com/article/10.1023/A:1021833110722
4. Mcleod, S.: Simply psychology (2014). Retrieved Winter 2018. https://www.simplypsychology.org/
5. Wilmshurst, J.: Measuring advertising effectiveness I: theories of how advertising works. In: Fundamentals of Advertising, pp. 316–336 (1999). https://doi.org/10.1016/b978-0-7506-1562-4.50021-0
6. Kenton, W.: Hierarchy-of-Effects Theory (2018). Retrieved Winter 2018. https://www.investopedia.com/terms/h/hierarchy-of-effects-theory.asp
7. Berger, A.A.: Media Research Techniques. Sage Publications, Thousand Oaks (1998)
8. Hoogs, A., Rittscher, J., Stein, G., Schmiederer, J.: Video content annotation using visual analysis and a large semantic knowledgebase. In: 2003 IEEE Computer Society Conference on Computer Vision and Pattern Recognition, Proceedings (2003). https://doi.org/10.1109/cvpr.2003.1211487
9. Pan, S.: The role of TV commercial visuals in forming memorable and impressive destination images. J. Travel Res. **50**(2), 171–185 (2009). https://doi.org/10.1177/0047287509355325
10. SealGrup, S.R.: Gerovital GH3 Men Shampoo – Advertising (2015). Retrieved Winter 2018. https://www.youtube.com/watch?v=yFGD21poKD8
11. ELITE: DASHING ELITE Men Shampoo - "Sticky Pole Men" (ENG) (2017). Retrieved Winter 2018. https://www.youtube.com/watch?v=d_oKFMfsSKM
12. Commercialfree: SHOPLIST18 年CM 色石篇 30秒 MEGASALE (2018). Retrieved Winter 2018. https://www.youtube.com/watch?v=X0S1QfhY-bU&fbclid=IwAR2mtlbi-I_TRe-qfktVrPMPsdwlZnfB925FgjDsXhOWFCKS78Mvzk1pRf
13. Oricon: 川栄李奈, 加藤諒演じる"ミス・チョイス"を撃退 美容脱毛サロン『銀座カラー』新CM「ミス・チョイスの罠」「ミス・チョイスを退治」 (2019). Retrieved Winter 2018. https://www.youtube.com/watch?v=BiiCrbtUYkI

14. TM Advertising: Japan Airlines - Perfectly Prepared (2015). Retrieved Winter 2018. https://www.youtube.com/watch?v=DDIkilSbxLQ
15. Vietnam Airlines: Vietnam Airlines - Reach Further - Proud to be 4 star Airlines - TVC 30 s (2017). Retrieved Winter 2018. https://www.youtube.com/watch?v=zUl1oQwSOGk

A Comparative Study on the Language Expressions of Cultural and Creative Products in the East and West

Chen Cheng[1], Junnan Ye[2], Hui Gao[1(✉)], and Guixiang Wu[1]

[1] School of Art and Design, Shanghai Institute of Technology, No. 120 Caobao Road, Xuhui District, Shanghai 200235, China
caca_hyuk@163.com, applegailgao@163.com, 2272188312@qq.com

[2] School of Art Design and Media ECUST, M.BOX 286, No. 130 Meilong Road, Xuhui District, Shanghai 200237, China
2723241@qq.com

Abstract. The concept of cultural and creative industries was first proposed by the United Kingdom. After that, UNESCO defined the cultural and creative industries. In recent years, China has proposed a series of new ideas and opinions on the innovation and development of cultural and creative industries, more and more experts and scholars have proposed to guide the development of China's cultural and creative industries by studying the cultural and creative industries of Western countries. However, the difference between China and the Western countries is not only reflected in the differences in the cultural heritage of the two sides, but also in people's understanding and expression of the respective cultures. For cultural and creative products, the cultural base is the design of the soul, then when how to transfer cultural concepts to product design in a subtle manner, the method is especially important.

In the paper, first of all, literature search, field research, expert interviews and other methods are used to find differences in lifestyle and state, cultural understanding and expression between China and Italy, representative of the Western countries. Through the KJ method and cluster analysis method, the cultural design language expressions of the forms, structures, functions, colors, human factors, materials and techniques of cultural and creative products in China and Western countries are summarized and compared. The design thinking and methods that can be used in the development of Chinese cultural products are obtained. Finally, taking China's Songxi culture and creative products as an example, the design practice has been carried out and has obtained good social and user evaluation. The research results will have important theoretical and practical significance for the design and development of Chinese cultural and creative products.

Keywords: Cultural and creative product · East and west · Language expression · Product design

A. G. Ho (Ed.): AHFE 2019, AISC 974, pp. 174–181, 2020.
https://doi.org/10.1007/978-3-030-20500-3_18

1 The Concepts of Cultural and Creative Industry and Its Products

Creative Industry is an emerging Industry that advocates innovation and individual creativity, emphasizes the support and promotion of culture and art to the economy under the background of knowledge economy and globalization. It is generally believed that as a national industrial policy and strategy, the creative industry was first proposed in Britain. In 1998, "The British Creative Industry Path" issued by UK Creative Industrial Task Force explicitly proposed the concept of "creative industry" and defined it as "an industry with the potential to create wealth and increase employment through the generation and utilization of intellectual property rights derived from individual creativity, skills and talents". This definition clarifies that the core content of the creative industry is culture and creativity, and emphasizes the driving force of culture and art to the economy.

And as "father of the creative industries," said the British economist Hawkins (John Howkins) published in his 2002 book "creative economy", and the creative industries as one sector of the economy and its products were within the scope of protection of intellectual property law, that copyright, patent, trademark and design industry, these four parts formed the creative industries and creative economy. This definition emphasized the important role of intellectual property in the creative industry, and includes patent research and development activities in the scope of creative industry.

The creative industry has a certain cultural features in nature, and the contents it carries need to be rich in cultural connotation. The support and promotion of culture and art to the economy are externally reflected in the creative industry. Creative Industry in content, as it were, and what we usually said the Cultural Industry has a lot of the same. The United Nations educational, scientific and Cultural Organization (UNESCO) in 2004, in the Montreal Meeting of "Cultural Industry", explained the definition as "the cultural industry is in accordance with industry standards, production, reproduction, storage and distribution of cultural products and services in a series of activities. From the definition of cultural products industry standard production, circulation, distribution and consumption perspective", and it can be divided into three parts, goods, services and intellectual property rights.

In China Creative Industry Development Report (2006), China Creative Industry Research Center defined creative industry as "the sum of those activities with certain cultural connotation, which are derived from human creativity and intelligence, and can be industrialized through the support of science and technology and market operation". Due to its cultural connotation, the creative industry is also called "cultural and creative industry" in some places [1].

The definition of cultural and creative products was put forward by Chen Zekai, head of the Manners in China-Ben Shi Jue Cultural and Creative Team, in the article "Culture of Taking Away – the definition and classification of cultural and creative products and 3c resonance principle". He defined cultural and creative products as products with market value derived from cultural themes and transformed through creativity. Generally speaking, cultural and creative products are understood as high value-added products produced by relying on the wisdom, skills, talents and cultural accumulation of creative people, creating and improving cultural resources and cultural goods, developing and applying intellectual property rights and by means of modern scientific and technological methods. The essence of cultural creative products is to emphasize a certain lifestyle and state that are representative of culture, and the taste that creative design improves for life. Such products integrate use and culture as a whole. When consumers buy them, they bring home a tangible cultural and creative commodity and an intangible culture at the same time [2].

2 Development Status of Cultural and Creative Industries in China

Modern China has transformed from a manufacturing power to a smart one. The Nation has put forward a series of new ideas and opinions on the innovation and development of cultural and creative industry, and it has become one of the hottest topics in China. Compared with the United Kingdom, the United States and other western countries, the concept of China's cultural and creative industry was proposed relatively late, and the concept of cultural and creative products is relatively vague, so many experts and scholars advised to guide the development of China's cultural and creative industry by studying the cultural and creative industry of western countries. But when you think about it, the differences between China and western countries are not only reflected in the differences in their respective cultural deposits, but also in people's understanding and expression of their respective cultures. For cultural and creative products, the basis of culture is the soul of design. How to instill the concept of culture into product design is particularly important. Only through correct understanding and understanding of the different ways in which westerners and easterners express their feelings and cultures in the design of cultural and creative products, can we find an effective way to develop China's cultural and creative industry, and increase our understanding of each other's cultural ethics and characteristics of people in such understanding and communication.

2.1 Interpretation of the Design Language of Western Cultural and Creative Products Represented by "Alessi Italy"

Alessi is a dream factory of Italian product design. With its unique shape, rich colors and changeable materials, combined with exquisite technology, it is full of poetic meaning. almost every shape of Alessi products has its own strong symbol concept, that is, unique but changeable. They range from alien-like lemon juicers (Fig. 1) to cartoon-like parrot bottle opener (Fig. 2) to "singing" kettles like precious works of art. In addition to the unique material, Alessi also carried out a bold attempt to color, bright and varied colors to increase the fun of many products. Marx once said, "the feeling of color is the most popular form of general beauty. Colour is the very important factor that affects consumer psychology. Alessi is multi-purpose in the colour of the product. Bright-coloured and colorful color will show household products lively, cheer the feeling with vigour, its main purpose still lets consumer be in actually the invariable in the most daily life that occupy the home, add a bit fun but no longer drab. Except color Alessi also think more about the material of products, stainless steel, thermoplastic resin, plastic is applied to product design skillfully, whether it is a single material or a combination of a variety of materials, designers have the characteristics of the material by the amount of the performance of clever, use it to express the design concept and implication of the each product [3].

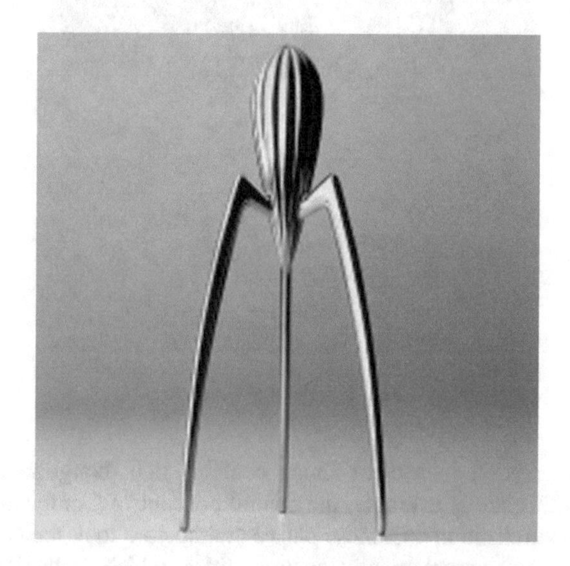

Fig. 1. Alessi Juicy Salif

As the most representative Italian cultural and creative product design, Alessi product's design is full of interest and human kindness. It makes each product into a work of art with "humanized" charm. Such cultural and creative products regard design as a manifestation of lifestyle, and focus on creating an interesting and affectionate lifestyle through products. In Alessi's cultural and creative products, the potential emotional needs of consumers are discovered and explored, and more attention is paid to the fun, humanization and interaction of products, so as to closely combine culture with people's personal life and create cultural and creative products with fun and topics to be played with [4].

2.2 Expression of Unique Chinese Cultural Forms in Product Design

In China, although the concept of "cultural and creative products" was proposed relatively late and has not been developed for so long, China has its own long cultural history and has formed some unique forms of cultural expression in its long history. Chinese traditional culture is extensive and profound and has a long history, which has created a rich material civilization for future generations.

Fig. 2. Alessi 9093kettle

The idea of creation in ancient China contains rich thoughts. Taking furniture design in ancient China as example, the central concept of Confucian aesthetics was "the beauty of neutralization". Chinese palace architecture took the central axis as the center and spread out symmetrically on both sides, which embodied the beauty of neutralization of Confucian culture. Chinese classical furniture was composed of wooden frame structure of archaize architecture, and its form of symmetry and harmony was derived from it. Confucianism attaches great importance to "the humanity", stressed "human relations", and the idea of "the way of the Holy King" was used by the ruling class. Therefore, in the feudal society, the imperial power was supreme, and there were distinct hierarchies and male superiority and female inferiority. This was reflected somewhat also on the use of furniture, design thought and aesthetic. Just as ji (a kind of table) was a symbol of hierarchy in the Han dynasty, emperors used jade table and princes used wood or bamboo table. Therefore, in terms of production and aesthetics, the object of use should be considered, which is related to the social and historical background at that time. Only with such a historical background can there be corresponding aesthetic appeal and aesthetic thought. In these ancient Chinese furniture products, not only reflects the traditional Chinese cultural thoughts, but also reflects the unique Chinese symbolic culture from the details of the carving. People put all sorts of good yearning contained in the carve patterns or designs of these furniture, used good stories, legends, or Chineseunique "homophonic" will behave the implied meaning of the products and indicatives. The form of pattern is characterized by auspicious folk customs, traditional symbolic techniques and rich spiritual connotation.

The content of carved pattern of Chinese ancient furniture has a few directions commonly. 1. A variety of historical figures and opera legend figures, generally symbolizing the educational music. For example, some opera characters, mostly based on the traditional drama of one or more scenes to form a pattern, make people to aesthetic education life, entertaining, four seasons peace, or good weather."Eight immortals crossing the sea" or "longevity" are all people's favorite representatives of the blessing. 2. A variety of playing pattern of dragon, phoenix and lion, a symbol of auspicious and evil. 3. All kinds of livestock cattle, horses, deer and other graphics, a symbol of fortune and good luck. 4. All kinds of poultry and lucky birds' graphics, a symbol of peace. 5. A variety of fish, insects, flowers and flowers patterns and their extensive use, the expression of happiness. 6. A variety of traditional artifacts and ancient treasures formed by the figure, meaning to ward off bad luck and welcome good luck. In addition, there are Fu Lu Shou Xi text and auspicious messages. China also has its own homophonic culture. For example, people are fond of carving "bats" in furniture, and westerners who cannot understand why they carve such ugly animals growing in dark corners into furniture. In fact, because in Chinese "bat" and "fu" have "blessing" homophonic, so in the folk meaning of life happiness. (Fig. 3) Generally with auspicious graphics, a bat means long life, two bats means happiness, three bats means healthy, four bats means good virtue. Therefore, in the design of these cultural artifacts, the story and the desire for a better life contained in each product are often more important than the actual use of the product itself [5].

Fig. 3. Carving "bats" in furniture

From the analysis of the formation of Chinese and western cultural products, it is not difficult to see that in the eyes of different people in the east and the west, there are significant differences in their cultural understanding. In the western cultural, creative products represented by Alessi, the product language it applies is more about creating an interesting and relaxing atmosphere. Colors, materials and shapes are mostly focused on relaxing and interesting features. What is more important is the appeal of the product, which brings people a more interesting and relaxed lifestyle. This is closely related to the western environment, social background, education and people's living habits. People prefer to live a light and interesting life. And in contrast, China's cultural products, people prefer to put own emotions into products, practical function is the least of its forming reasons, but more reflect the identity and the honour, such as different people or the people with the fear of ghosts-gods legends, or is the use of all kinds of people, animals, objects form semantics to reflect people's yearning for a better life. It is not difficult to see that the Chinese heart is reserved, introverted and life is always full of beautiful expectations of the characters. The cultural foundation of China is more profound. Only by effectively exploring the cultural connotation of China and expressing it in the form of Chinese people's understanding, can we truly design our own cultural and creative products [6].

3 New Ideas of Chinese Cultural and Creative Products

Danner is a French historian and literary critic. In "philosophy of art", Danner analyzed a large number of historical facts, deeply discussed some typical cultural phenomena, analyzed and compared them, and scientifically revealed the close relationship between culture, art and the three elements of race, environment and era. Among the three factors of race, environment and era, race is the internal, environment is the external pressure, and era is the acquired driving force [7]. It is the interaction of these three factors that influences and restricts the development and trend of spiritual culture including culture and art. Only by truly understanding and analyzing the three factors applicable to China itself can all cultural creativity be promoted effectively.

In 2018, China held a regional cultural promotion project named "thousand-year-old Songxi and hundred-year-old sugarcane". Songxi county, as the first national ecological county in northern Fujian, has a long history, a long history and a distinctive region. Songxi county has experienced the process of germination, origin, formation, development and prosperity after thousands of years. It absorbs the essence of ancient Yue culture, central plain culture and Confucianism, Taoism and Buddhism culture. We will promote Songxi county with a long history and cultural characteristics as an experimental cultural creative product design exploration. We applied this interpretation method of Chinese culture to the cultural and creative product design of Songxi, and let some new generation of designers conduct some souvenir design through their own investigation and re-analysis of Chinese culture and Songxi regional culture, and got good feedback. Not only the local people and local leaders are very much appreciated and loved, experts and ordinary tourists from all over the world, have given highly comments on the concept of such souvenirs. (Fig. 4) the designer uses the Taoist thought and auspicious "Hui" pattern in the traditional Chinese culture to skillfully design the tea set. The product color matching technology and the pattern of his flower table have the unique antique flavor of Chinese style, and their delicacy coexists with the atmosphere.

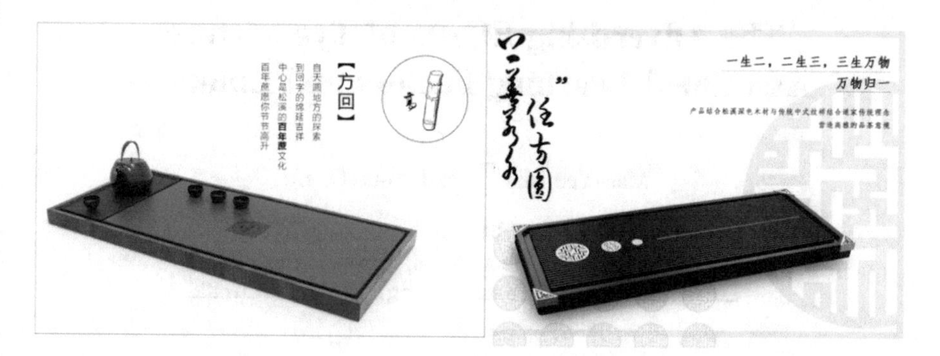

Fig. 4. The cultural and creative product design of Songxi

4 Conclusion

By comparing the design languages of cultural and creative products of China and the west, we have enhanced our understanding of each other's culture and lifestyle, enhanced our understanding of each other and enhanced the opportunity of cultural exchange between China and the west. We also have a deeper understanding of China's own national characteristics and cultural ethics and have deeper feelings. On the basis of mutual understanding, China can certainly step out of its own more comprehensive road of cultural and creative product design.

References

1. Ling, J., Zhang, X.: Present conditions and research of China's creative design industry. Creat. Des. 22–39 (2012)
2. Chen, Z.: Take away culture-definition and classification of creative products and "3C resonance principle". Mod. Commun. 103–105 (2017)
3. Ge, C.: Analysis and transformation of requirement in the design process of cultural creativity product. Art Des. 142–143 (2018)
4. Vijay, K.: 101 Design Methods_A Structured Approach for Driving Innovation in Your Organization. Wiley, Hoboken (2012)
5. Wanhong, Z., Hongliang, S., Yan, W.: The Traditional Culture of China. Beijing Normal University Publishing Group, Beijing (2012)
6. Cagan, J., Vogel, C.M.: Creating Breakthrough Products - Innovation from Product Planning to Program Approval. China Machine Press, Beijing (2004)
7. Taine, H.A.: Philosophy of Art. Modern Press Co., Ltd. (2017)

The Advertising Effects of Typotecture: Associated Learning Factors and Emotions

Man-Yee Mak(✉) and Amic G. Ho

Department of Creative Arts, The Open University of Hong Kong, Jockey Club Campus, Ho Man Tin, Kowloon, Hong Kong
maxenemak@gmail.com, amicgho@ouhk.edu.hk

Abstract. Typography undoubtedly plays a critical part in print advertisings. While most of the typographies we encountered in advertisings are two-dimensional, typotectures are types that hybridise architecture and typography together. A remarkable feature of typotecture is that its semantic language directly combines with the semiotics of the visual sign. This paper aims at discussing the possible benefits of using typotecture on print advertisings. In order to do so, six representational typotecture advertisings will be used to conduct studies and to find out what perceived experiences and emotions can audience perceive in the typotectural designs.

Keywords: Typotecture · Advertising · Typotectural advertising · Emotion design · Associated learning factors

1 Introduction

The word 'typotecture' means types being able to subject themselves to gravity and acquire physical presences, to expand themselves into space and come closer to architectural forms. Typotecture had first been brought to attentions in the 1910s and 20s. Much attention had then been drawn to Typotecture in 1927. Futuristic graphic artist, Fortunato Depero, transferred the expressive quality of letters into the ephemeral architecture like the exhibition building, Padiglione del line, in 1927, in Monza. Regarding on posters, Miháky Biró, a Hungarian by birth, who later worked in Vienna and Berlin, was one of the first to recognise the potential of typotecture and to explore it in an apparently systematic way [1] (Tables 1 and 2).

Typotecture is chosen for this research because of its easiness in eliciting power and emotions on advertising material like posters and book covers. As mentioned by Felix Studinka, the expansion of types into architectural forms develops powers and emotional sensitivity that posters have always yearned for [1].

Another reason typotecture is being the subject of this study is because very few researches have been done regarding how typotecture posters can be used in advertisings including print advertisings to deliver wanted emotions and meanings to audience. This research paper also wants to find out if typotecture can evoke perceptual experiences of a product, location, movie and service, etc., to audience. Ultimately, it aims to find out if typotecture can persuade audience to get involved with the advertising. Hence, such a study is worth to be done.

A. G. Ho (Ed.): AHFE 2019, AISC 974, pp. 182–193, 2020.
https://doi.org/10.1007/978-3-030-20500-3_19

2 Important Findings of Typotecture

Jander suggested that typotecture could be categorised into three categories based on Peirce's theory of signs [1]. Iconic sign is an iconic image conveys a message with its resemblance to the signified object. An indexical sign is an indexical image indicates the causal relationship between the image and the reality. The symbolic sign is a symbolic image represents meanings of the reality based on social norms, customs, and contexts [2, 3]. Janser pointed out that the significant property of typotecture is that the semantics of the language is directly combined with the semiotics of the visual sign. A typotecture image is intended to be looked as much as to be read. This iconic dovetailing suggested that semantics and semiotic elements reinforce each other mutually [1]. In other words, when investigating the effect of typotecture on printed advertisings, it is crucial to find out both what is being seen and hidden and how they reinforce each other to fit the message of the advertising.

Emotions on Typotecture Advertising Designs. One of the perspectives about emotion is that it is a feeling state that constitutes cognitive, behavioural, and physiological reactions to internal and external events [4].

Emotion is considered a special nature interacting with our actions, values and judgments. Feelings allow emotions to affect our mind, and when without emotions we could not get amusement in life or to appreciate what around us [5]. Ultimately, emotions give meanings and significance to things surround us [6]. Hence, it is important to consider emotions when discussing the benefits of using typotecture in advertising designs.

The reason emotion needed to be included when discussing the advertising effect of typotecture imageries is because earlier researches have shown the intertwined relationship between persuasion and emotions. People in positive emotion are more susceptible to persuasion in typotectures [7]. People view the world through rose-coloured glasses in a good mood. Quicker decision making could be done as they are unlikely to review information already examined, but rely more on heuristic cues [8, 9].

The 'feelings-as-information' view also suggested that while positive moods signal to people that everything is fine and no effortful thought is necessary, negative moods signal that something is wrong in their environment and that some action is necessary [10]. Eventually, people in a positive mood are more susceptible to persuasion as they are less likely to engage in extensive thinking of the presented arguments than those in a neutral or negative mood.

Recent Research conducted by Owolabi [11] also suggested that a positively evaluated context evokes positive moods, which leads to positive advert evaluation and lesser advert processing. This phenomenon is attributed to the fact that a positive mood reduces the processing of stimulus information.

3 Theoretical Framework

Perceiving Information. While sights are formed through eyes, visions are formed within our brains. In the cortex of the brain, visual information takes either one of the two routes. Thalamo amygdala-pathway is a crude network of swift pre-conscious processing. Signals are sent immediately to Thalamus. "Neurological research reveals that visuals may be processed and form the basis of future action without passing through consciousness at all" [5]. "The second route is the slower one. Visual information is sent to the amygdala (located deep within the temporal lobe), where emotion is attached to incoming data and a feeling is generated. When an individual senses something visually in the environment, the brain attaches emotion. The signal is sent across the neural networks, beginning in the left hemisphere, stretching across to the right hemisphere, where the brain attaches the experience to a radiating network of other semantic connections [12]."

Associated Learning Factors. Associated learning factors affect users experience on types. Prior research has found out that the reader reacts to the text, here, typography based on either learned arbitrary associations, figurative associations, or abstract associations [13]. "Learned arbitrary association refers to the influence of historical precedence on affective response to typography. Figurative association occurs when the typeface resembles something in the real world. Finally, the semantic association can be associated automatically causing abstract association [14]."

Emotion. "Emotion is a process of appraisal of anger, disgust, fear, happiness, sadness, and surprise. The experience of emotion is individual, social, and cultural. Feeling proceeds from the initial emotion appraisal and affect the manifestation of the feeling. The feeling is a behavioural response (neurological and physical) to the primary emotion states (anger, fear, joy, disgust, sadness, shame, and guilt) [12]."

Although no studies have combined vision and emotion processes with typotecture design, the significant study on responses to emotionally weighted pictorial stimuli and neurological imaging of conscious visual attention have been examined. Evidence for an impressive capacity of parallel affective discrimination in rapid picture presentation was found in the study of affective picture processing by Peyk, Schupp, Keil, Elbert, and Junghofer in 2009 [15]. Their findings show that subjects were able to discriminate pleasant from neutral image content. They further concluded, "The emotion discrimination at the level of perceptual processing is a robust phenomenon" [15].

This shows that any emotionally pictorial stimuli have an effect on human emotion. Since typotecture images also contain certain degree of emotionally pictorial stimuli and, therefore, it also has an effect on human emotion. Hence, the theoretical framework, based on a paper published by Shaikh in 2007, should be outlined as follows. When a print typotectural advertising is being perceived by the audience, association learning factors such as learned arbitrary association, figurative association, and, or semantic association is induced. It is essential to stress the concept of association learning factors does not contradict with Peirce's theory of signs. Since both iconic signs and figurative associations represent typeface resembles something in the real world. Both indexical signs and learned arbitrary associations represent the causal relation between typeface and reality. In addition, last both symbolic signs and abstract associations represent deeper and more abstract meanings.

The next step followed by finding out what semantic and emotion meanings can the audience get out of these associations. Emodi [16] opines that "Semantics means the study of meaning. Semantics deals basically with the mind to give appropriate meaning to a word or an expression."

Finally, to see if the typotectural posters are appropriate and persuasive by checking if the perceived semantic and emotional meanings of the typotecture match with the context of the advertising.

4 Research Methodology and Results

4.1 Research Methodology

Hypotheses. (1) Typotectural posters project higher intensity of visual impact than typographic posters. (2) Participants prefer typotecture imagery posters to typographic posters. (3) Typotectural posters are rich in associative learning factors - Learned arbitrary association, figurative association and semantic association [17]. These associative learning factors help the audience to create positive/appropriate emotions and semantic meanings of the advertising. Since positive emotions and appropriate meanings of an advertisement can increase the persuasiveness of an advertising [7]. Therefore, typotectural posters can be used as a tool to enrich the persuasiveness of advertising.

Study 1. In the first study, participants were asked to fill in questionnaires regarding on 6 representational typotectural posters and 6 modified versions of these posters in witch the typotectures are replaced by types that do not exhibit any architectural forms. Each participant was given a link to login to an online questionnaire created by using SurveyHero to ensure they can take a clear look at all of the posters

Subjects. 100 online questionnaires were sent out to participants aged 18–64 and 83 questionnaires were received with completion. Out of the 83 returned questionnaires, 71.08% of them were female and 28.92% of them were male.

Materials. 6 representational posters and book covers from different categories including films, magazine, book, bookshop and arty shopping centre were used in this study. 6 control imageries were created to suit the purpose of this study by replacing the typotectural designs into typographic designs (Figs. 7, 8, 9, 10, 11 and 12).

Fig. 1. PMQ Poster Retrieved from http://www.pmq.org.hk/event/pmq-tourist-privilege-programme-coupon-book-redemption-jun18/?lang=ch

Fig. 2. TimeOut Cover Retrieved from https://www.debutart.com/artist/stephan-walter/made-in-london-time-out

Fig. 3. TimeOut Cover Retrieved from http://creative-lab.co.uk/blog/2016/05/02/designing-gerard-woodwards-legoland-cover-picador/

Fig. 4. Manhattan Poster Retrieved from https://www.pinterest.com/pin/195836283767885743/?lp=true

Fig. 5. Eslite BookstorePoster Retrieved from https://i.pinimg.com/originals/b1/ba/3f/b1ba3f3e50b65edfa30e3c4ba9712efd.jpg

Fig. 6. West Side Story Poster Retrieved from https://www.1stdibs.com/furniture/wall-decorations/posters/west-side-tory-film-poster-1961/id-f_7384313/

Fig. 7. PMQ Modified version

Fig. 8. TimeOut Modified version

Fig. 9. Legoland Modified version

Data Analysis and Results. Majority of participants favour typotectural designs over typographic designs. Over 90% of the participants prefer typotecture design to typography design for both Legoland and PMQ print advertisings. More than 80% of the participants favour typotecture designs over typographic ones for the Dream Big poster and TimeOut Magazine cover. For Manhattan film poster, 75.90% prefer typotectural version to the typographic version. For the West Side Story film poster, 67.47% of participants prefer the typotectural version to the typographic version. Hence, participants prefer typotectural posters to typographic posters in this study.

Fig. 10. Modified Manhattan Poster

Fig. 11. Modified Big Dream Poster

Fig. 12. West Side Story Modified version

Based on the below result, the average visual impact of these typotectural posters/book cover is 3.66, which shows that participants do find these typotectural designs catch their attention. Amongst all the typotectural designs, the Legoland book cover and PMQ Poster have the highest visual impact mean values, 4.16 and 4.11 respectively, which could due to the exquisiteness of these typotectural designs. Detailed 3-dimensional types, which appear as swimming pool, roller coaster, tower, outdoor cinema, etc., are found in the Legoland book cover. Similarly, 3-dimensional types appear as the shopping mall, food building and experimental building are found in the PMQ poster. These architecture-form types could help stimulate the audience's imaginations and allow them to experience the story of the advertising. Hence, participants favour typotectural designs over typographic designs.

On the contrary, West Side Stories and Manhattan posters have the lowest visual impact mean values, 3.05 and 3.04 respectively, which could be the reason that these typotectural designs are in 2-dimensional types instead of 3-dimensional. On top of that, both the architectural elements in these posters are relatively simpler comparing to the Legoland and PMQ ones.

Table 1. Means and Standard Deviations of each typotectural posters or imageries visual impact. 5 = Maximum & 1 = Minimum

Poster/Imageries	Mean	SD
Legoland	4.16	0.41
PMQ	4.11	0.40
TimeOut	3.81	0.38
The Eslite Bookstore	3.42	0.38
Manhattan	3.40	0.38
West Side Story	3.05	0.43
Average	3.66	

Study 2. In the second study, 12 participants were asked to conduct in-depth interviews regarding the typotectural posters assigned to them. 1-on-1 interviews are conducted by a trained researcher, with the role is to deliver questions and to give appropriate guidance to ensure participants to finish their interviews smoothly. Each typotectural poster/book cover was being used for both 2 males and 2 females to conduct the interviews. The purpose of the study is to find out what kinds of association learning factors and emotions can participants get out of these typotectural designs. Hence, to find out if participants would be persuaded by the advertisements and take actions in involving with the ads.

Subjects. 12 participants aged 18–35 were invited to participate in the interviews, with 6 males and 6 females.

Materials. The 6 typotectural posters and book covers were used in this study. The Six Basic Emotion picture developed by The Grimace Project and The Feeling Wheel developed by Dr Gloria Willcox are also as tools to facilitate the interviews. Six Basic Emotions is a theory developed by American Psychologist Paul Ekman and Wallace V. Friesen. They believe that Sadness, Joy, Surprise, Anger, Fear and Disgust are 6 universal and fundamental emotions for humans [9].

4.2 Results of Typotecture Evoke Learned Arbitrary Associations

Legoland. When asking the participants what kinds of personal/perceptual experience (s) they received when viewing the typotectural design in Fig. 3. One of the participants said the letter D swimming pool reminded him of his childhood swimming experience he had at home. The movie screen on the letter N reminded him of the experience of driving his car into one of these outdoor movie theatres. Another participant claimed that the billboard/screen on the letter N reminded her of people going on holidays to get rid of hectic work lives.

TimeOut. Participants claimed that the TimeOut typotecture, Fig. 2, reminded them of the experience of going to music shows, musicals, carnivals and art districts. One participant even claimed that it reminded her of her school art crafting projects.

West Side Story. Most of the participants did not think the typotecture, in Fig. 6, reminded them of any personal experiences. Only one participant thought that the stairs on the scruffy types reminded him of going to a ghetto place for travelling. Nonetheless, some did agree that there is some perceptual experience perceived, such as people chasing in these stairs located New York angrily.

PMQ. Most of the participants thought that Fig. 1 it reminded them of the experience of going to a festival, carnival or a place where there are interesting things to see, great food to eat and nice things to shop. Some of the participants think that the typotecture reminded them of doll houses, toy sets and miniature cities they have played when they were young.

The Eslite Bookstore. While most participants could not think of any perceptual or personal experiences due to its abstract of Fig. 5. However, one participant claimed that it reminded her of looking at an architecture blueprint or some sort of architecture sketches.

Manhattan. Most of the participants think that Fig. 4 reminded them of going to metropolitan cities like Manhattan or skyscraper areas in Manhattan.

Since the descriptions towards each typotecture design are based on the participants' historical precedence. In other words, events they have experience or learnt before. Hence, the typotectures from Figs. 1, 2, 3, 4, 5 and 6 showed evidence of evoking learned arbitrary associations for participants.

4.3 Results of Typotecture Evoke Learned Arbitrary Associations

For the Legoland typotecture, Fig. 3, participants claimed that they could see the swimming pool, movie screen, roller coaster, cafe, shopping mall and daily life objects there. For the TimeOut magazine typotecture, Fig. 2, people claimed that they could think of showbiz logos, number plates, entertainment logos, broadway logos. For the West Side Story typotecture, Fig. 6, participants claimed that they could think of wall paintings, bricks, and stairs in public housings, escape stairs and stairs in Rosenheim, Germany. For PMQ typotecture, Fig. 1, it is obvious that it reminded participants of PMQ or some sort of all-in-one arty shopping centre. Other thought included generic modern art museum. For the Manhattan typotecture, Fig. 4, participants tended to think it simply reminded them of skyscrapers in Manhattan or other metropolitan cities.

For the Eslite Bookstore poster, Fig. 5, most of the participants cannot find any figurative associations to it due to its abstract. Nonetheless, one participant's answers were hitting the head of a nail. She claimed that the typotecture reminded her of the blueprint of an architectural project and the types looked like sketching buildings on a floor-plan design.

4.4 Results of Typotecture Evokes Semantic Association

Since the semantic association is caused by semantic properties being associated automatically by reviewers towards a type. It is an immediate abstract association of type instead of the actual semantic meaning of the text. The semantic association is retrieved from semantic memory. Semantic associations are random facts and associations, which could include information that is encyclopaedic in nature, lexical or language related, and conceptual [14].

When considering random facts and association participants have towards the book cover Legoland. Participants described it as an interesting anecdote, a book tells the interesting operation of Legoland and entertainment about leisure daily lives. Another conceptual idea a participant voiced is that the helicopter on the typotecture design gives the idea of freedom.

For the TimeOut typotecture, some conceptual ideas are raised such as new ideas being built by a group of young producers to create new things for people to see and the idea of patriotism, pride and quality by the participants. For the West Side Story typotecture, random associations such as a movie in an urban city, low-class story, romance in the city, a thriller with dancing elements, west side story, ghetto were raised by the participants. For PMQ typotecture, random facts and associations such as modern art, exhibition, festival, carnival, people, public drink, food and art were raised by the participants. For the Eslite Bookstore typotecture, random facts and associations such as creativity, unlimited possibilities, imaginations, sleek and modern design were raised by the participants. For the Manhattan typotecture, random ideas such as drowning in buildings, metropolitan cities, comedy, thriller, suspend, romance were raised by the participants.

Perceived Emotions on Each of the Typotecture Design. The aforementioned result shows that typotecture did indeed convey association-learning factors to participants. The important question then to ask is what other emotions and feelings can participants get out of these typotecture design since emotions help the audience to make sense and give the meaning of products and furthermore to persuade the audience to take action in involving with the advertising [11].

Table 2. Means of emotions perceived and their intensity regarding each of the posters.

Categories	Joy	Surprise	Sadness	Anger	Disgust	Fear
Legoland	3.00	4.00				
TimeOut	4.00	2.00				
West Side Story	2.00	0.50	0.50	1.50	1.00	1.00
PMQ	4.00	4.00				
The Eslite Bookstore	1.50	1.50				
Manhattan	2.00	2.33				

5 Interests in Getting Involve with the Advertising

Most participants found the messages of the advertising for PMQ, TimeOut, Lego-land, West Side Story matches with its typotecture design. When asked them about the theme of these typotecture advertising, they showed little to no difficulties in answering the questions.

Moreover, most of the participants showed interested in purchasing or getting involved with these advertisements. For example, most participants showed interest in reading the book or finding out more information about the book Legoland and the TimeOut magazine after the interview. In addition, most participants showed interested in going to PMQ for entertainment and leisure purposes.

This matches with the above results regarding the emotions perceived by participants. Most participants received a moderate to high level of joy and surprise when seeing these typotecture designs and hence they felt it more appealing to get involved with the advertising. Furthermore, the aforementioned results also showed participants could perceive an abundant amount of associated learning factors from these typotectures. Therefore, these typotecture posters are rich in emotional and semantic meanings.

Above half of the participants, consider watching the movie Manhattan. The reason participants claimed would not consider watching is that they think that the movie seemed out-dated; the colour of the design is too plain. Another thing that is worth noticing is that the typotecture elements of this typotecture design are not as prominent as others.

Although the emotional values and typotecture elements of the West Side Story typotecture are relatively lower than its counterparts, people still showed interests in watching the movie. One of the possible explanation is because the associated learn-ing factors evoked from the advertising gives important meanings and information for the participants be persuaded to get involved with the movie.

Most of the participants showed no interests in going to the Eslite Bookstore. That matches our results findings, as most of the people found that the typotecture did not evoke many emotions and associated learning factors. Hence, participants find it hard to understand the message of the advertising, some even said it was abstract to understand and they can find out the theme of the advertising. In terms of the typotecture elements of the design, though it is a 3-dimensional typotecture. However, there lack figurative associations for this design. Therefore, participants might found it too abstract to get to the message of the advertising and lost interest in it.

6 Conclusion

In conclusion, studies show that typotecture designs do give a significant visual impact for participants and participants do favour typotecture advertising designs over typotecture designs. On top of that due to the rich in associated learning factors perceived from typotecture designed, it could help advertising gives emotions, semantic meanings and users' experiences. Hence, significant benefits are shown by using typotecture in advertising materials to persuade consumers. Although, a few of the results from study 2 were not very positive regarding on the leaned association factors people get from the typotectural advertisings. Still, the overall results showed promising potential of typotecture in advertisings. Due to its power in evoking past or perceptual experience, advertisings related but not listed to events, museums, shows, exhibitions could consider putting typotecture elements into their advertising materials. In the end, the potential of typotectural is unlimited and the purpose of the paper is raise the awareness of typotecture and be creative with it.

Acknowledgements. I want to give thanks to my professor Dr Ho Amic Garfield for his support and guidance. Thanks for guiding me through the process of writing this thesis paper.

Dedication. I would like to dedicate this thesis to my dad, whose soul is resting peacefully now. Dad! You taught me once as a girl that I should equip myself with knowledge and walk through life with courage. Thank you for introducing me to reading, travelling and arts.

References

1. Janser, A.: Typotecture: typography as architectural imagery = Typotektur: Typografie als architektonische bilderwelt. Museum für Gestaltung Zürich, Plakatsammlung, Zürich, Switzerland (2002)
2. Panksepp, J.: At the interface of the affective, behavioral, and cognitive neurosciences: decoding the emotional feelings of the brain. Brain Cogn. **52**, 4–14 (2003)
3. Izard, C.E.: Emotion Theory and Research: Highlights, Unanswered Questions, and Emerging Issues. Annu. Rev. Psychol. **60**, 1–39 (2009)
4. Carlson, J.G., Hatfield, E.: Psychology of Emotion. Harcourt Brace Jovanovich, San Diego (1992)
5. Barry, A.: Perception theory. In: Smith, K., Moriarty, S., Barbatsis, G., Kenney, K. (eds.) Handbook of Visual Communication: Theory, Methods, and Media, pp. 45–62. Lawrence Erlbaum, Mahwah (2005)
6. Gross, J.J.: Emotion regulation: past, present, future. Cogn. Emot. **13**(5), 551–573 (1999)
7. Janis, I.L.: Facilitating effects of eating – while-reading on responsiveness to persuasive communications. J. Pers. Soc. Psychol. **1**, 181–186 (1965)
8. Schwarz, N., Bless, H., Bohner, G.: Mood and persuasive communications. In: Zanna, M. (ed.) Theories in Experimental Social Psychology, vol. 24, pp. 161–199. Academy Press, San Diego (1991)
9. Handel, S.: Classification of Emotions, 24 May 2011. Retrieved 30 April 2012
10. Schwarz, N.: Feelings as information: information and motivational functions of affective states. In: Higgins, E.T. (ed.) Handbook of Motivation and Cognition: Formations of Social Behaviour, vol. 2, pp. 527–561. Guilford Press, New York (1990)

11. Owolabi, A.B.: Effect of consumers mood on advertising effectiveness. Europe's J. Psychol. 118–127. (2009)
12. Koch, B.E.: Human emotion response to typographic design. Ph.D. thesis, University of Minnesota (2011)
13. Doyle, J.R., Bottomley, P.A.: Dressed for the occasion: Font-product congruity in the perception of logotype. J. Consum. Psychol. **16**(2), 112–123 (2006)
14. Shaikh, A.D.: Psychology of onscreen type: investigations regarding typeface personality, appropriateness, and impact on document perception. Ph.D. thesis, Department of Psychology, College of Liberal Arts and Sciences, Wichita State University (2007)
15. Peyk, P., Schupp, H.T., Keil, A., Elbert, T., Junghöfer, M.: Parallel processing of affective visual stimuli. Psychophysiology **46**(1), 200–208 (2009). https://doi.org/10.1111/j.1469-8986.2008.00755.x
16. Emodi, L.N.: A semantic analysis of the language of advertising. Afr. Res. Rev. **5** (2011). https://doi.org/10.4314/afrrev.v5i4.69286
17. Eid, M., Larsen, R. (eds.): The Science of Subjective Well-Being: A Tribute to Ed Diener, pp. 44–61. Guilford Publications, New York (2008). ISBN 978-1-59385-581-9
18. Curralo, A.F., Soares, L.: Pedagogical encounters: typography and emotion. In: Proceedings of the 3rd International Conference for Design Education Researchers, p. 698 (2015)
19. DeSteno, D., Wegener, D.T., Petty, R.E., Rucker, D.D., Braverman, J.: Discrete emotions and persuasion: the role of emotion-induced expectancies [abstract]. J. Pers. Soc. Psychol. **86**, 43 (2004)
20. Gardner, M.: Mood states and consumer behaviour: a critical review. J. Consum. Behav. **12**, 281–300 (2004)
21. Lyubomirsky, S., Sheldon, K.M., Schkade, D.: Pursuing happiness: the architecture of sustainable change. Rev. Gener. Psychol. **9**(2), 111–131 (2005)
22. Mac Luhan, M., Fiore, Q.: The Media is the Massage. An Inventory of Effects. Gingko Press Inc., Corte Madera (2001)
23. Collins, A.M., Loftus, E.: A spreading activation theory of semantic processing. Psychol. Rev. **82**, 407–428 (1975). https://doi.org/10.1037//0033-295x.82.6.407
24. Mizerski, R., White, D.: Understanding and using emotions in advertising. J. Consum. Mark. **3**, 57–69 (1986). https://doi.org/10.1108/eb008180
25. Petermans, A., Pohlmeyer, A:. Design for subjective well-being in interior architecture. In: Proceedings of the 6th Symposium of Architectural Research, Finland (2014)
26. Pohlmeyer, A.: Design for happiness. Interfaces **92**, 8–11 (2014)
27. Peirce, C.: Peirce on Signs: Writings on Semiotic by Charles Peirce. University of North Carolina Press, Chapel Hill (1991). Accessed 2012

Research on the Historical Context of Guangzhou Time-Honored Catering Brands' Narration

Xiaobao Yu(✉) and Qian Peng

Art & Design School, University of Shenzhen, Shenzhen, Guangdong, China
yxb99-1@126.com, 1017025303@qq.com

Abstract. The culture of Guangzhou food is extensive and profound for a long time. It contains the unique culture history of South five ridges. Importantly, its unique dining custom constitutes local traditional folk culture. Guangzhou time-honored restaurants update their way of brand narration, no matter in content or in form as they correspond to a alteration, such as the TaoTaoJu restaurant, LianXiangLou restaurant, Panxi restaurant and Guangzhou Jiujia restaurant. In fact, this reveals the historical context of Guangzhou's catering consumption culture from another perspective.

Keywords: Guangzhou time-honored catering brands · Narrative form · Narrative content

1 Introduction of this Article

With strong South of the five ridges cultural characteristics, Guangzhou time-honored catering brands are special business cards of Guangzhou city and they provides important recreation place for citizens. As a symbol of local culture, they contain rich brand culture and legendary brand stories. However, with the changes in the consumption context of the times, these brands are confronted with new fierce challenges and competition. Scholars have made admirable breakthroughs in terms of brand's value enhancement, brand's image transformation, and revival of time-honored brands. Regretfully, they have not systematically sorted out and analyzed the historical context of the time-honored brand's narrative. Therefore, we still face troubles in understanding their development history and practical dilemma, not to mention their protection and inheritance. Based on the theory of brand narrtology, this article will sort out the historical context of Guangzhou time-honored catering brands, thus providing a strong brand culture support for the revival of the time-honored catering brands.

The research on historical context of Guangzhou time-honored catering brands' narration will start with the trace of brands, including its occurrence, development and evolution in a theory system. The main content of the article will explain when the research object started to have brand narrative behavior. What are the forms of brand's narration? What are the main contents of the narrative and what information are they trying to convey? The narrative of the local Time-honored catering brand started in a specific regional culture. Therefore, before answering these questions, it is necessary to

A. G. Ho (Ed.): AHFE 2019, AISC 974, pp. 194–203, 2020.
https://doi.org/10.1007/978-3-030-20500-3_20

sort out the history of Guangzhou catering culture considering its historical context of analysis comprehensively by combining those factors such as culture, technology and media of different ages. We do not intend to sort out the characteristics of the time-honored brand objectively in the past 100 years of historical evolution, but clarify the context of these characteristics.

2 Historical Evolution and Characteristics of Guangzhou Catering Culture

The catering culture is gradually formed by the influence of multiple factors such as history, economy and politics. The formation of Guangzhou's catering culture is also the same. Maba The earliest ancient people in south of the five ridges area have already known how to use natural fire to cook food, that mark the human diet into a state of civilization. On the basis of the use of fire, the local catering culture in the Pre-Qin period began to flourish. Archaeologists discovered that our ancients had already domesticated livestock and found the remains of wild animals in northern Guangzhou. It indicates that in the early days of human civilization, Guangzhou catering culture has a wide range of ingredients.

Partial ancestors of Guangdong found a suitable place to live three or four thousand years ago and developed into a clan tribe and a tribal alliance. The South Vietnamese who had a large number of population began to localize after the Western Han Dynasty. In the process of gradual integration, Cantonese cuisine began to mainly accept the eating habits of Nanyue people. Until the development of the sugar industry in the Han Dynasty, Cantonese cuisine was seasoned with sugar and developed into a characteristic of advocating fresh.

The speed of Cantonese cuisine growing rapidly during the Tang Dynasty. After the completion of the integration of Han and Yue, those factors include the stable political and economic environment, adequate food supply and excellent cookware design in Tang Dynasty gradually made citizens' higher requirements for food. In turn, this promoted the cooking skills of Cantonese cuisine maturely and perfectly. According to historical records, in the Tang Dynasty, there were already many cooking methods such as boiling, roasting, deep-frying, steaming, stir-frying, braising, pan-frying simmering, mixing, etc. The prototype of the cooking technique system has gradually formed. Due to the wide range of ingredients, Cantonese cuisine has begun to explore its own cooking style based on different ingredients. And the category of Cantonese cuisine changes constantly because of the high tolerance of foreign things. Then developed into it's own classification.

During the Ming and Qing Dynasties, Guangdong's agriculture develop rapidly thanks to the advances in water conservancy construction and farming techniques, and the enthusiasm for food of people continued to increase after they settled down. At the same time, the tendency of 'talking of drinking and eating' in Guangzhou has become increasingly popular, many famous Guangzhou cuisines have gradually appeared in the choices of people's dining. And influenced by political factors, the Ming government Promulgating the sea ban policy, there only three cities set up the city's shipping department. Guangzhou is the only one which connected to Southeast Asia and

Western countries. Quanzhou just led to Ryukyu, and Ningbo just connected to Japan, thus Guangzhou established a foundation for the development of trade. The looting of Western colonialism during the Qing Dynasty led the government to shut down the city's shipping divisions in Ningbo and Quanzhou, remain the operation of Guangzhou, it further stimulating the prosperity of Guangzhou's economy. At this stage, Guangzhou merchants gathered here, the culture of cooking developed rapidly. With the introduction of Western food, local food are made by Guangzhou chefs based on massive absorption of foreign cooking techniques. Cantonese cuisine is booming and its status in the food culture is increasing day by day.

The position of Guangzhou in foreign trade was even more important after the Opium War, the cultural and economy of Guangzhou was prosperous at the same time, and the merchants around the world gathered here, further driving the demand of the catering industry. At the end of the Qing Dynasty and the beginning of the Republic of China, Guangzhou catering presented a scene of prosperity, it has continued into the 1930 s, 'Eating in Guangzhou' has been well-known at domestic and foreign. According to records, there are more than 200 large-scale catering stores in Guangzhou, and they all have their own signature dishes. The development of Cantonese cuisine has entered a stage of prosperous.

Thanks to the introduction of our reform and opening policy, special overseas trade status of Guangzhou makes it easier to integrate the essence of different food cultures between Central and western, and getting more innovative. The Guangzhou Municipal Government has hosted the Guangzhou International Food Festival every year since 1987, and other cities in Guangdong also hold events with the theme of 'food culture'. After the 1980s, Cantonese cuisine began to be exported to areas outside, eat Cantonese food become more and more popular, and it make its mark in Chinese cuisine.

By sorting out of the Guangzhou's catering brands' history development, we made a form of it which divide into 4 parts to reveal the period, historical stage, catering cultural characteristics and internal motivation of development (Table 1).

Table 1. Overview of Guangzhou catering history development and characters.

Period	Historical phases	Characters	Motivation of development
Pre-qin	Sprouts of catering	Wide range of ingredients	Hard living conditions
Han & Tang	Forming cooking system	Sweet taste; variety of cooking methods; cooking form is depending on the materials	Stable political and economic environment; adequate food supply and excellent cookware design
Ming & Qing	Rapid growth period	Extensive absorption of foreign cooking techniques based on local food making	The establishment of the city's shipping division; the influx of Western food culture

(*continued*)

Table 1. *(continued)*

Period	Historical phases	Characters	Motivation of development
Republic of China	Prosperous period	Choose materials and production, more catering brand have built; and they have their own signature dishes	Foreign trade booming and the demands of business activities
Contemporary	Stationary period	Hosting an international food festival	Outputting the food culture

3 The Historical Context of Guangzhou Time-Honored Catering Brands' Narrative

In the early stage of brand development, the popularization of brand-name products can't be classified as mature brand narrative behavior due to the deficiency of brand culture consciousness, but it accumulated rich experience and materials for the further brand building. Brand personality and brand culture will be built consciously when a company has a branding conception. The way to enhance customer relationships effectively and gain brand value will follow, and also the behavior of brand narrative will matured from the initial.

In order to investigate the historical context of the Guangzhou Time-honored Catering brands' narrative comprehensively and delicately, this paper consulted lots of historical texts and conducted on-the-spot investigations on the research objects. With the objective analysis of brand narrative form and content in their development history, the development of narrative forms are divided into three stages by the means of spreading. And the development of narrative contents are divided into five stages by the demand of audience.

3.1 Historical Evolution of Brand Narrative Forms

The geographical resources of Guangzhou are always going to be more advantaged as well as the south Five-Ridges catering culture accumulated in the past, these brands established an important foundation for their development. The prosperity of Guangzhou's business during the Ming and Qing Dynasties promoted many catering brands. With the development of the media, these brands continue to accumulate their own brand resources by the power of brand narrative, thus establish a stronger brand image.

The Brand Narrative Form in the Spreading Era 1.0. Early advertising industry in China is not developed due to they were penetrated with a conventional idea that "emphasized agriculture and limited trade", and the spreading channels of brand are very few. The brand information is disseminated only rely on those traditional ways such as sipping, selling, displaying, plaques, and postings. The Time-honored brand-

TaoTaoJu, founded by Huang Chengbo in the Qing Guangxu six years (1880), the reason why scholars gathered here is that the elegant environment of TaoTaoJu, varieties of tea and where the location of it is in the center of city, and the people who came here for business negotiations and wedding meetings with the same reason. TaoTaoju attract more and more customers here by seeking antithetical couplet and displaying tea utensils and cakes. Based on the theme “Tao Tao” to seeking antithetical couplet and sets different levels of bonuses. scholars come to the sound. The reputation of TaoTaoju has spread in range for a time. In addition, Tao Taoju focused on the theme of tea, and displayed the tools and the horses that used to get spring water for brewing tea on the lively street deliberately. it directly indicate that quality water come from the mountain that is not easy to get. This activity has been widely spread as a popular news, The form of brand narrative at this stage mainly carries out brand public relations activities with the brand story, relying on the audience to spread by creating brand hot news.

The Brand Narrative Form in the Spreading Era 2.0. After the Opium War in 1840, Europe countries tried to establish the Chinese market to dump goods, They failed due to China’s own self-sufficient small-holder economic model. Under this circumstance western companies attempted gain the attention of Chinese consumers by advertisements such as posting trademarks, painting walls, and delivering colored cards. Influenced by Western advertising and modern printing, the domestic business newspaper industry began to gradually expand its advertising business in the 1860 s and 1990 s. The appearance of business newspapers and advertising companies has made brand communication awareness active. This stage is the coexistence period of brand word-of-mouth communication and newspaper advertisement communication. From the “report” published in 1884 until the “Guangzhou news” in 1886, and the “China-Spain Daily” in 1891, those all reserve space for advertising. The “Chinese-Western Daily” published types of advertisements are more than the previous newspapers, including some catering teahouse pubs. TaoTaoju advertised on newspaper in order to establish the cake market, in addition to recycled pottery moon cake boxes and sole moon cakes with gifts. In the 1980 s, as China implements the reform and opening up policy, the creativity and activity of enterprises were greatly stimulated, and the types of advertisements were increasingly enriched on the market. Besides newspaper advertisements, there are window advertisements, street signs advertisements, photography, etc. And TV advertisements have grown rapidly as a medium of extraordinary performance at the time. As a new media tool, it complete the connection between people and things (people) through images without staying at the scene, which greatly alleviates problems such as the information sharing obstacles in spreading, or the delay of urgent communication information and the information asymmetry in spreading. And TV can simultaneously the feelings of sight and hearing of people, it makes TV advertising become one of the main narrative forms of brand’s for a long time. However, some precious image materials are hard to be found again due to the low preservation of television materials (Fig. 1).

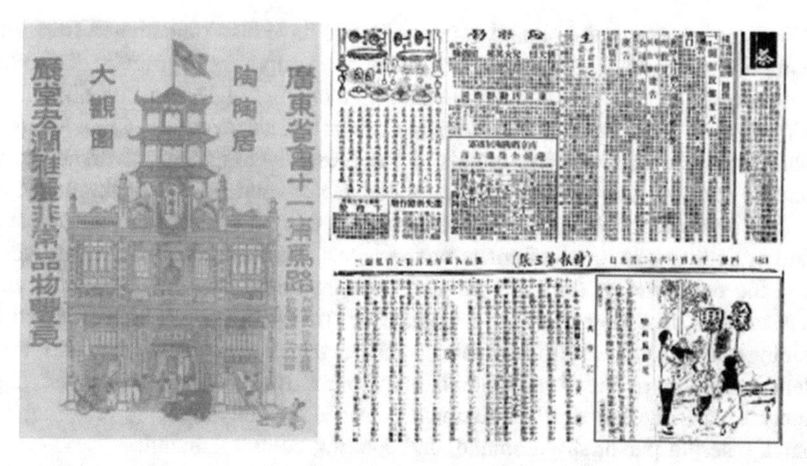

Fig. 1. Part of newspaper advertisements of Taotaoju restaurant. And images from the web and full-text database of the period of the late Qing and the Republic of China.

The Brand Narrative Form in the Spreading Era 3.0. With the invented and development of Internet, the spreading channels of brand's communication are no longer limited to these traditional media, and new forms of brand narrative have emerged. The mode of traditional media is too monotonous and lack of interaction with the audience as the carrier of the brand narrative form. It is another milestone in brand narrative that brand take new media as a carrier of its narrative form, not only the form become diversity but also can interact better with the audience. Through the advantages of three-dimensional communication of new media, Guangzhou time-honored catering brands have created their own self-media accounts to spread information on different Internet platforms such as WeChat, Weibo and Taobao. Enterprises interact with audience through vivid and specific brand stories, delivering the brand culture while deepening the brand perception of the audience. For example, Lianxianglou restaurant write an article on Wechat titled "I'm sorry, but we did something that made Guangzhou people proud about 129 years ago", it tells us a story of master Chen Chenwei who made the high-quality lotus seed with the golden color by using text form In 1889, and this made Lianxianglou's cake became the pioneer of the Cantonese-style lotus seed cake, and attraced more customers to popularize Lianxianglou. Audience acquire some brand information automatically through this WeChat tweets such as the brand market status, brand history and brand culture. In addition, the company answers audience questions and interacts with them in the comment area, further deepening the intimacy between audience and brand.

4 Content Evolution of Brand Narrative

With the development of times, people pay attention to the goods from its material function to the spiritual function. The exchange of products or services has evolved from the initial material exchange to a symbol exchange, with the changes of audience, brand narrative content is evolving constantly as a carrier of symbol.

Nowadays, these time-honored catering brands have also transitioned from simple consumption of catering function to a complex consumption of symbolic. This change can be reflected in the changes from the narrative content of Guangzhou time-honored catering brands. From the day when Huang Chengbo, Chen Weiqing, Chen Xinghai, Li Wenlun establish their catering brands to the present, the change of brand narrative content affected by the political factors, social factors and factors of the times, it's a shift that from material to spiritual, essentially, this is a performance indicate the change of audience in terms of consumption levels and their consumption attitudes. In theory, the evolution of the time-honored catering brand narrative content is in line with Maslow's hierarchy of needs it pointed out that people's needs are ranked from physiological demand, security demand, social demand to respectful demand, and the content evolution of the time-honored brand narrative evolved from satisfying the audience's catering function requirements, catering quality demand, catering social demand, catering personality demand, and catering culture demand.

The Demands of Catering Function. In the early, the narrative content of the time-honored catering brands mainly to addresses the physiological needs of consumers for food. Guangzhou people's diet chart includes breakfast, morning tea, lunch, afternoon tea, dinner, night tea, staying up late, and they start to eating from 5am to 3am until the morning in the next day. Under such circumstance, Guangzhou's tea house restaurant has gradually developed to facilitate people's dietary needs. In addition, restaurants offer ceremonial grounds such as funeral, weddings and some other festivals. At this stage, these time-honored catering brands mainly focused on the function of the store and the promotion of food in the narrative content. The content of Guangzhou Jiujia restaurant was published in the first volume of the "style of Weinan" in 1948 that introduced the food types that they provide and they have the ability to undertake kind of feast, as the picture shown below (Fig. 2).

Fig. 2. This figure of Guangzhou Jiujia restaurant that was advertised on the newspaper before, it indicate that the brand will change their narrative content as the demands of customers are changing. Image come from the web and full-text database of the period of the late Qing and the Republic of China.

The Quality Demands of Food. The narrative content of the time-honored catering brands focus on the consumer's need for food quality. Naturally people always choose foods that with better taste and quality. The merchants will do their best on the dishes because of that. the selection of food, the cooking skill and taste of food will be more reflected within the narrative. For example, TaoTaoju did their best on the designing and matching of dishes to attract more customers. It has created the "Chinese food be eaten in a west way", "the most distinctive 'full meal', is divided into two different sizes, but each with same crab meat shark's fin and four dishes of in-seasons vegetable, fried rice or a bowl of fish thornback noodle, fruits, melon seeds, and then ended with the tea, it's popular in the market because it's cheap and convenient." [1] on the selection of materials, Lianxianglou emphasizes the promotion of its quality lotus seed fillings made of high quality lotus seeds and high quality white sugar. With the food production technique, the comging of "Kohl Niu Liu" is an interesting story of Panxi Restaurant. When German Chancellor Kohl went to Guangzhou to discuss the subway project in November 1993, he canceled the arranged banquet, and going to the Panxi Restaurant, he refused to enter the VIP room arranged by the hotel insisted on sitting in the hall and ordered several Cantonese dishes. Suddenly, he proposed to eat one kind of sour dish that made by beef, onions, peppers and fruits, the chefs present the "Yakyu NiuLiu" After a while, Prime Minister Cole greatly appreciated that after eating, and later this story was spread to customers, the name of this dish was changed to "Kohl NiuLiu" in the end.

The Social Demands of Dietary. The narrative content of the time-honored catering brands values the demands of consumers to maintain social relationships. Scholars have proposed that "social catering network has two functions that are emotion and instrumentality" [2]. That is to say strengthen relationship and making new friends. catering store provides an environment for the audience to interact with each other, customers can consolidate and deepen their social relationships which already existed, while gaining new social relationships in this scenario. The narrative content of time-honored catering brands is gradually approaching to satisfy the needs of social function within catering. the book "Famous Tea House Taotaoju" which was awritten by Feng Mingquan, it mentioned that the hall where TaoTaoju was set up on the third floor was not only the first choice for the scholars those people want to avoid being disturbed, but also for those people who bought the concubine to avoid being seen. And since the 1960 s, the Panxi Restaurant has ability that receive honoured guest for domestic and foreign maturely, including former British Prime Minister Heath, former US President George W. Bush, former German President Kohl, Deng Xiaoping, and Zhu De. A group of photo was published on the "Guangzhou University Journal" in 1948, indicate that Guangzhou Jiujia restaurants has the ability to satisfy the space needs of high-end business gatherings.

The Demands of Brand Personality. The narrative content of the time-honored catering brands is transferred to solve the consumer's demand for their identity and personality. "Consumers regard brands as partners, friends and even self extensions of themselves, they entrust the personality traits to brand and choose brands like choose friends naturally" [3]. The time-honored catering brands can satisfy some consumers' personality demand from intrinsic value to appearance. Firstly, the appearance of the brand is the thing that consumers can directly perceive from the senses, such as brand design, interior decoration, tableware design. The design of the Panxi Restaurant is the most distinctive which designed and hosted by the famous garden architecture design expert Mo Bozhi. The garden restaurant with large-scale landscape is full of infinite wisdom without any words to explain. Brand intrinsic value is the highlight of it's own personality. Guangzhou people advocates working hard, believes in spirit of innovation and enjoys their lives. The time-honored catering integrates these characteristics into themselves, one of the initial characteristics is to invest a lot of energy in food production to impress customers through the foods, hard working make brand more stable in the market, and the innovative spirit of brand has been demonstrated in enhancing the sensory experience of the diners when they come. The garden of the Panxi Restaurant is the example that perfectly indicates the brand which knows how to enjoy life.

The Demands of Brand Culture. The narrative content of the time-honored catering brands is more and more concerned about how to solve the consumer's demand for inner spirit. "The establishment of brand culture is the highest level, which is in line with the self-realization needs of consumers. It is also the essence of the company's establishment of a century-old brand. Brand culture is to satisfy the high-level spiritual needs of consumers" [4]. The old brand has become a microcosm of the local food culture after years, its value has long exceeded people's basic needs for food, as a vivid symbol of the history and culture for Guangdong, memories of citizens has been carried on with it. A local resident of Guangzhou once proudly said that he used to eat Taotaoju since he was a little kid, the food was tasty there. On the one hand, the locals choose to consuming in time-honored catering brands means that they have commendation with the local food culture, on the other hand, it indicates that they have clear self-identification with confidence in their geographical identity. Many strangers choose these brands because they represented the catering culture of Guangzhou. We could observe that the trend of their culture has been adjusted and developed in the matured narrative content. In the Mid-Autumn Festival in 2018, the moon cake advertisement titled "I always thought that parents do not like to eat salted egg yolk" which launched by Guangzhou Jiujia Restaurant tells a story about a man who has gained a deeper understanding of family and memories from university to society. It's mainly emphasizes that "love and reunion", but it also reflected that the moon cakes of Guangzhou Restaurant are an important link for family reunion from the other side, further highlighting the cultural value of the time-honored brand.

5 Conclusion

While a brand has a theme to indicates the core value of itself, it will take various resources to tell the story in different ways. With the rapidly changing in information, the form of brand narrative and the story deserve to be valued today. They are both interrelated in brand development, affecting the brand's core concept and brand culture. Especially for the time-honored catering brands which need to exhibit new brand culture to the audience by telling a story, and establish their own brand values through brand narrative today. On the way to promote the brand's progress, time-honored catering brands needs relevant leader who have a keen insight to find a balance between the brand and the audience to establish a close relationship in the future, in order to find the audience to narrative accurately.

References

1. Tan, L., He J.F.: Guangzhou Time-Honored Brand, Guangzhou Culture History广州文史. Guangdong People's Publishing House, Guangzhou (2001)
2. Wang, X.L.: Food in Guangzhou: Lingnan catering Classic Culture食在广州 岭南饮食文化经典. Guangdong Tour Press, Guangzhou (2006)
3. Wu, Q.: China's Academic History of Advertising (1815–1949)中国广告学术史 (1815–1949). Intellectual Property Publishing House, Beijing (2014)
4. Full-text database of journals in the late Qing Dynasty and the Republic of China. http://www.lib.szu.edu.cn/er/bksy-wqmg
5. Dacheng old publication full-text database. http://www.lib.szu.edu.cn/er/dacheng
6. Bian Y.J.: Comparison of China, Japan and Korea in the social function of food and beverage network餐饮网社交功能的中日韩比较. J. Academic exchange. **155** (2013)
7. Xu Wei.: Brand personality, identity and loyalty of the old brand–development and evaluation of personality scale老字号品牌个性、认同与忠诚——个性量表开发与评价. J. Financ. Theory **4**, 95–96 (2013)
8. Bian Y.J.: Viewing Enterprise Brand Building from Maslow's Demand Theory从马斯洛需求理论看企业品牌建设. J. Times Econ. **6**, 32 (2008)

Author Index

A. G. Ho (Ed.): AHFE 2019, AISC 974, p. 205, 2020.
https://doi.org/10.1007/978-3-030-20500-3

Printed in the United States
By Bookmasters